Lecture Notes in Physics

Edited by H. Araki, Kyoto, J. Ehlers, München, K. Hepp, Zürich
R. Kippenhahn, München, H. A. Weidenmüller, Heidelberg
J. Wess, Karlsruhe and J. Zittartz, Köln
Managing Editor: W. Beiglböck

290

K. T. Hecht

The Vector Coherent State Method and Its Application to Problems of Higher Symmetries

Springer-Verlag
Berlin Heidelberg GmbH

Author

K. T. Hecht
Physics Department, University of Michigan
Ann Arbor, MI 48109, USA

ISBN 978-3-662-13633-1   ISBN 978-3-540-48011-2 (eBook)
DOI 10.1007/978-3-540-48011-2

© Springer-Verlag Berlin Heidelberg 1987
Originally published by Springer-Verlag Berlin Heidelberg New York in 1987
Softcover reprint of the hardcover 1st edition 1987

2153/3140-543210

Acknowledgements

These notes are based on a series of lectures first given in the fall of 1986 at Niigata University, Niigata, Japan, repeated in part at Tokyo Metropolitan University, Tokyo, Japan, and given in final form in the spring 1987 at the Max-Planck-Institut für Kernphysik, Heidelberg. It is a pleasure to thank Prof. K. Ikeda and Niigata University, Prof. H. Toki and Tokyo Metropolitan University, and Prof. H.A. Weidenmüller and the Max-Planck-Institut für Kernphysik for their warm hospitality and the Japan Society for the Promotion of Science for its support. Special thanks go to Prof. Y. Suzuki of Niigata University for his encouragement which led to the writing of these notes and his many suggestions for their improvement.

Contents

1. Introduction.

Coherent states (and their many generalizations) are now used so widely in all branches of physics that the term "coherent state" has become a "household" phrase (and as such is used very loosely!). For an overview of how coherent state theory is used in many applications in physics, there are two recent books on coherent states, [J. R. Klauder and B. S. Skagerstam, "Coherent States", 1985; a reprint volume with an introductory chapter], and [A. Perelomov, "Generalized Coherent States", 1986]. One of the first articles on coherent states [by R. J. Glauber, 1963], dealing with quantum optics, (hence the origin of the name "coherent state"), is an excellent introduction to the subject.

The aim of these lectures, however, is a very limited one: to show how a generalization of coherent state theory, termed "vector coherent state theory", can be used to give a very explicit construction of the matrix representations and the Wigner-Racah calculus of many of the higher rank symmetry algebras, with applications particularly to nuclear physics problems, with some excursions into quark models. The vector coherent state theory is an outgrowth of the study of the nuclear collective model from a microscopic point of view, the Sp(6,R), [often also called Sp(3,R)] symmetry model.

The vector coherent state techniques are an outgrowth of $Sp(6, R)$ studies by three main groups, at

University of Toronto (D. J. Rowe and collaborators)

Université Libre de Bruxelles (C. Quesne and collaborators)

University of Mexico (M. Moshinsky and collaborators),

with somewhat different perspectives, language, and emphasis used by the different groups. The Mexico group uses the language and techniques of boson realizations, [Castaños et al.; 1984, 1985, 1986]. The generalization of standard coherent state theory to a theory of vector coherent states was introduced independently by the Toronto and Bruxelles groups, [Rowe 1984 a,b; Rowe, Rosensteel, and Gilmore, 1985; Rowe et al. 1987 preprint; Deenen and Quesne, 1984 a,b; and Quesne, 1986]. These lectures will follow very closely the method as expounded in Toronto; since it is the Toronto group which has carried out the most elegant and useful developments of this method and has made the most extensive applications. One of the central features of the vector coherent state method is the associated K-matrix theory. It is the Toronto group, in particular, which has developped the K-matrix technique into a viable computational tool which has led to the many recent detailed applications. A good introduction to the vector coherent state method is given by D. J. Rowe, ["Some Recent Advances in Coherent State Theory and its Applications to Nuclear Collective Motion", in Phase Space Approach to Nuclear Dynamics; 1986].

Coherent states were created in the context of the harmonic oscillator. The harmonic oscillator is extremely simple from the algebraic point of view. For more complicated problems the algebraic approach often becomes quite tedious. The vector coherent state method eliminates most of this complexity by reducing a large number of problems involving higher symmetries to problems of an n-dimensional harmonic oscillator algebra coupled with a simple intrinsic symmetry algebra which is of lower rank than the full symmetry algebra. The final solution of the problem involves some "vector" (or analagous higher symmetry)-coupling of the intrinsic algebra with the n-dimensional oscillator algebra. As a result, the matrix representations and the Wigner coefficients of the higher rank symmetry algebra can be expressed in terms of simple calculable functions and in terms of recoupling coefficients of the simpler intrinsic symmetry algebra.

Since the vector coherent states are a generalization of *one* of three different ways of defining coherent states, where these three different definitions are equivalent, (or essentially equivalent), for the 1-dimensional harmonic oscillator, but in general differ for problems of higher symmetry, we will first review the standard coherent state theory as defined in connection with the 1-dimensional harmonic oscillator.

1. 1   Standard Coherent States. The
1-dimensional Harmonic Oscillator.

If physical coordinate and momentum are measured in standard units, (such as $\sqrt{\hbar/m\omega}$ for $x_{physical}$ and $\sqrt{\hbar m\omega}$ for $p_{physical}$), the 1-dimensional oscillator hamiltonian, expressed in terms of dimensionless coordinate, x, and dimensionless momentum, p_x, is

$$H = \frac{1}{2}(p_x^2 + x^2) = \frac{1}{2}(a^\dagger a + a a^\dagger),$$

where this applies both to the motion of a particle of mass m, or to a mode, (say the kth one), of the radiation field; and where a, and $a^\dagger$ are given by

$$a = \frac{1}{\sqrt{2}}(x + \frac{\partial}{\partial x}), \qquad a^\dagger = \frac{1}{\sqrt{2}}(x - \frac{\partial}{\partial x}).$$

The simple algebra of these oscillator annihilation and creation operators, (the 1-dimensional Heisenberg-Weyl algebra), is

$$[a, a^\dagger] = 1, \qquad [a, 1] = 0, \qquad [a^\dagger, 1] = 0.$$

There are three types of coherent state definitions. For the 1-dimensional oscillator they are all equivalent (to within a c-number function). I will name them: I, the "geometrical" definition; II, the "generator function" definition, useful for the construction of a complete set of states, (and the one whose generalization will actually be used in these lectures; and III, the "coherent" state definition, (or "true coherent state" definition).

Starting with the geometrical definition (I), consider an arbitrary wave function $\psi(x)$ describing some system. Let the system be displaced through a distance, c. Then

$$\psi(x - c) = \sum_{n=0}^{\infty} \frac{(-c)^n}{n!} \frac{\partial^n}{\partial x^n} \psi(x) = exp(-c\frac{\partial}{\partial x})\psi(x) = e^{\frac{c}{\sqrt{2}}(a^\dagger - a)}\psi(x).$$

By defining $\frac{c}{\sqrt{2}}$ as a new quantity, z, and allowing z to be an arbitrary complex number, we make a generalization of the displacement operator. Letting this act on the oscillator vacuum, $|0\rangle$, yields the first definition of the coherent state. Definition I:

$$e^{(z^*a^\dagger - za)}|0\rangle = |z\rangle_I$$

Let the "displacement" operator $D(z) = e^{(z^*a^\dagger - za)}$. Then using the special case of the Baker-Campbell- Hausdorff relation

$$e^{(A+B)} = e^A e^B e^{-\frac{1}{2}[A,B]}$$

valid, when

$$[A,[A,B]] = 0, \qquad \text{and} \qquad [B,[A,B]] = 0,$$

this leads to

$$e^{(z^*a^\dagger - za)} = e^{z*a^\dagger} e^{-za} e^{-\frac{1}{2}zz^*}.$$

We get

$$D^{-1}(z)aD(z) = a + z^*; \qquad D^{-1}(z)a^\dagger D(z) = a^\dagger + z;$$

or

$$D^{-1}xD = D^{-1}\frac{(a + a\dagger)}{\sqrt{2}}D = x + \frac{(z^* + z)}{\sqrt{2}}$$

$$D^{-1}\frac{1}{i}\frac{\partial}{\partial x}D = D^{-1}\frac{(a - a^\dagger)}{i\sqrt{2}}D = \frac{1}{i}\frac{\partial}{\partial x} + \frac{(z^* - z)}{i\sqrt{2}}.$$

Thus, for $z = $ real number, $D(z)$ effects a shift in coordinate space, for $z = $ pure imaginary number, $D(z)$ effects a shift in momentum space; and in general is a shift operator in phase space. Using the Baker-Campbell-Hausdorff identity, we could also have expressed this coherent state by

$$|z\rangle = e^{z^*a^\dagger} e^{-\frac{1}{2}zz^*}|0\rangle = e^{-\frac{1}{2}zz^*} \sum_{n=0}^{\infty} \frac{(z^*)^n}{\sqrt{n!}}|n\rangle,$$

where $a^\dagger|n - 1\rangle = \sqrt{n}|n\rangle$. Except for the c-number function $e^{-\frac{1}{2}zz^*}$ this is equivalent to

Definition II

$$|z\rangle_{II} = e^{z^*a^\dagger}|0\rangle = \sum_{n=0}^{\infty} \frac{z^{*n}}{\sqrt{n!}}|n\rangle.$$

This may be the simpler definition if our interests lie in state-constructions, since it generates all oscillator excited states in the simplest way.

Note that the scalar product, (which follows from the orthonormality of the standard oscillator excited states, $|n\rangle$),

$$\langle z'|z\rangle = e^{z'z^*}$$

does *not* lead to δ-functions. (For definition I, we would have had $\langle z|z\rangle = 1$; but $\langle z'|z\rangle$ is again a function of z, z'). The coherent states are overcomplete. The type II coherent states lead to the expression of the 1-operator

$$1 = \frac{1}{\pi} \int d^2z \; e^{-zz^*}|z\rangle\langle z|$$

where the integration is over the full z-plane. To show that 1 is indeed a unit operator, introduce polar coordinates

$$z = \varrho e^{i\phi}$$

so that

$$\frac{1}{\pi} \int d^2z \; e^{-zz^*}|z\rangle\langle z| = \sum_{n,m} \frac{1}{\pi} \int d^2z \; e^{-zz^*} \frac{z^{*n}}{\sqrt{n!}}|n\rangle \frac{z^m}{\sqrt{m!}}\langle m|$$

$$= \frac{1}{\pi} \sum_{n,m} \int_0^{2\pi} d\phi e^{i(m-n)\phi} \int_0^{\infty} \varrho d\varrho e^{-\varrho^2} \frac{\varrho^{n+m}}{\sqrt{n!m!}}|n\rangle\langle m|$$

$$= \sum_{n,m} \delta_{nm} \frac{1}{n!} 2 \int_0^{\infty} d\varrho \varrho^{2n+1} e^{-\varrho^2}|n\rangle\langle n| = \sum_n |n\rangle\langle n|.$$

State vectors $|\psi\rangle$ can be expressed through their z-space functional realizations

$$\langle z|\psi\rangle = \psi(z).$$

This leads to the scalar products

$$\langle\psi_1|\psi_2\rangle = \frac{1}{\pi}\int d^2z\ e^{-zz^*}\langle\psi_1|z\rangle\langle z|\psi_2\rangle = \frac{1}{\pi}\int d^2z\ e^{-zz^*}\psi_1^*(z)\psi_2(z).$$

The z-space is the so-called Bargmann space, the z-space functions, $\psi(z)$, are the Bargmann transforms of the ordinary coordinate representations, $\psi(x)$. Note that

$$\psi(z) = \langle z|\psi\rangle = \sum_n \frac{z^n}{\sqrt{n!}}\langle n|\psi\rangle = \sum_n \frac{z^n}{\sqrt{n!}}c_n.$$

These $\psi(z)$ could also have been obtained directly from the coordinate functions $\psi(x)$ by the Bargmann transformation with the kernel function, $A(x,z)$,

$$A(x,z) = \frac{1}{\pi^{\frac{1}{4}}}\exp(-\frac{1}{2}x^2 + \sqrt{2}xz - \frac{1}{2}z^2),$$

with $$\psi(z) = \int\limits_{-\infty}^{+\infty} dx A(x,z)\psi(x).$$

Using the generating function for Hermite polynomials

$$e^{-s^2+2sx} = \sum_n H_n(x)\frac{s^n}{n!}, \qquad \text{with} \qquad s = \frac{z}{\sqrt{2}},$$

$$\psi(z) = \sum_n \int\limits_{-\infty}^{+\infty} dx\frac{e^{-\frac{1}{2}x^2}H_n(x)}{\sqrt{2^n n!\sqrt{\pi}}}\frac{z^n}{\sqrt{n!}}\psi(x) = \sum_n \frac{z^n}{\sqrt{n!}}c_n.$$

Note, in particular, if $\psi(x)$ is an oscillator eigenstate, say $\psi(x) = \langle x|n\rangle$, then its Bargmann transform is $\frac{z^n}{\sqrt{n!}}$. The $\psi(z)$ are entire functions, (analytic everywhere in the finite z-plane). In the z-space realization, with scalar products defined in terms of the z-space integrals with the Gaussian Bargmann measure, e^{-zz^*}, the

$$\chi_n(z) \equiv \frac{z^n}{\sqrt{n!}}$$

are the orthonormal basis functions.

Not only state vectors, but also operators acting on such vectors, can be transformed into their z-space realizations. If

$$|\psi\rangle \Longrightarrow \psi(z) = \langle z|\psi\rangle = \langle 0|e^{za}|\psi\rangle$$

operators, O, map into

$$O|\psi\rangle \Longrightarrow \Gamma(O(z, \frac{\partial}{\partial z}))\psi(z) = \langle z|O|\psi\rangle = \langle 0|e^{za}O|\psi\rangle$$

$$= \langle 0|(e^{za}Oe^{-za})e^{za}|\psi\rangle = \langle 0|\left(O + [za, O] + \frac{1}{2}[za, [za, O]] + \cdots\right)e^{za}|\psi\rangle.$$

Since operators can be built from oscillator annihilation and creation operators, it will be sufficient to determine the z-space realizations of a and $a^\dagger$. For a only the first term in the commutator expansion survives, and the a can be brought down by differentiating the exponential with respect to z. For the case, $O = a^\dagger$, note that $a^\dagger$ acting to the left on $\langle 0|$ gives zero, while the first (single) commutator term gives z. Thus

$$a \Longrightarrow \Gamma(a) = \frac{\partial}{\partial z}; \qquad a^\dagger \Longrightarrow \Gamma(a^\dagger) = z$$

Note that these z-space realizations of the operators satisfy the proper commutation relations. Note also that $\partial/\partial z$ is the adjoint of z with respect to a scalar product defined with the Bargmann measure. If $\psi_a(z)$ and $\psi_b(z)$ are arbitrary analytic functions given in terms of the expansions

$$\psi_a = \sum_{n=0} a_n z^n; \qquad \psi_b = \sum_{n=0} b_n z^n;$$

$$\frac{1}{\pi} \int d^2z \, e^{-zz^*} \psi_a^* \frac{\partial}{\partial z} \psi_b = \sum_{n=0} a_n^* b_{n+1} (n+1)! = \frac{1}{\pi} \int d^2z \, e^{-zz^*} \psi_b [z\psi_a]^*.$$

Both coherent states of type I and type II satisfy another property: They are eigenstates of the oscillator annihilation operator, a. This leads to

Definition III

$$a|z\rangle = z^*|z\rangle.$$

This property follows at once, (using definition type II),

$$a|z\rangle = a \sum_{n=0} \frac{(z^*)^n}{\sqrt{n!}} |n\rangle = \sum_{n=1} \frac{(z^*)^n}{\sqrt{(n-1)!}} |n-1\rangle$$

$$= z^* \sum_{n=1} \frac{(z^*)^{n-1}}{\sqrt{(n-1)!}} |n-1\rangle = z^*|z\rangle.$$

We should here make a small footnote: It is this property of the oscillator coherent state in its applications to quantum optics which gave rise to the name "coherent state". In applications to problems of optical coherence, [see Glauber; 1963], it is useful to split the electric field operator into a "positive frequency", photon absorption part, and a "negative frequency", photon emission part,

$$\mathbf{E} = \mathbf{E}^{(+)}(\mathbf{r}, t) + \mathbf{E}^{(-)}(\mathbf{r}, t),$$

where

$$\mathbf{E}^{(+)} = \sum_k \varepsilon_k(\mathbf{r}, t) a_k,$$

and where a_k is the photon annihilation operator for the kth mode and $\mathbf{E}^{(-)}$ is the hermitian conjugate of $\mathbf{E}^{(+)}$. The coherence of the electromagnetic field is specified by the 1,...,n-photon correlation functions. The 1-photon correlation function, *eg*, is

$$G^{(1)}(\mathbf{r},t;\mathbf{r}',t') = trace\big(\varrho E^{(-)}(\mathbf{r},t)E^{(+)}(\mathbf{r}',t')\big)$$

and involves the absorption of a photon at space-time point $(\mathbf{r}',t')$ and the emission at $(\mathbf{r},t)$. If the density operator, ϱ, which describes the state of the field , is given in terms of a single coherent state, the normalized correlation functions of the above type have their maximum possible value of unity, specifying the optically coherent state.

Although the type III definition, (the "true coherent state definition"), is equivalent to definitions I and II for the simple oscillator, this equivalence breaks down in almost all generalizations of this standard coherent state. In our applications generalizations of type II coherent states will be most useful. In that sense then the word "coherent state" is somewhat of a misnomer.

Other Simple Examples:

Before generalizing the coherent state to the so-called "vector" coherent states, let us look at some simple generalizations of the standard coherent states, involving other simple symmetry algebras. Again we start with the geometrical point of view. The oscillator coherent states were related to translations in phase space. Clearly rotations and dilations should also be considered. The rotations are of course related to the angular momentum algebra. If we attempt the correspondence

$$a^\dagger \to J_+; \qquad a \to J_-; \qquad 1 \to J_0;$$

where the J_i satisfy the angular momentum algebra, then the geometrical, or type I, generalization of the coherent state might be

$$e^{z^* J_+ - z J_-}|M_{min.} = -J\rangle,$$

where the oscillator vacuum has been replaced by the angular momentum eigenstate with the minimum possible M-value, which is annihilated by J_-. Again z is to be an arbitrary complex number, but if we express this in terms of the two real numbers θ and ϕ, by

$$z = \frac{1}{2}\theta e^{i\phi}$$

then

$$e^{z^* J_+ - z J_-} = e^{-i\theta(J_x \sin\phi - J_y \cos\phi)} = e^{-i\theta(\mathbf{J}\cdot\mathbf{n})},$$

where $\mathbf{n}$ is a unit vector in the x, y-plane making an angle ϕ with the negative y-axis. This is illustrated in Fig. 1. Thus, by proper choice of θ and ϕ this coherent state is associated with rotated systems. (The fact that rotations about the z-axis seem to be excluded is not crucial for the physics. If the z-axis is the quantization axis, a rotation about the z-axis merely induces a change of phase of an angular momentum eigenstate).

To attain one more simple generalized coherent state, associated with still another algebra, consider the dilation of a system. For a 1-dimensional system described by $\psi(x)$, the dilated system follows from the operation

$$exp(-\alpha x \frac{\partial}{\partial x})\psi(x) = \psi(e^{-\alpha}x).$$

Note that this follows from a power series expansion of $\psi(x)$ and the identity

$$(x\frac{\partial}{\partial x})x^m = m x^m, \qquad e^{-\alpha x \frac{\partial}{\partial x}}(x^m) = (e^{-\alpha}x)^m.$$

For a system of N particles in 3 dimensions, the dilation operator can be generalized to

$$exp(-\alpha \sum_{s=1}^{N} \sum_{i=1}^{3} x_{si} \frac{\partial}{\partial x_{si}}),$$

where

$$a_{si} = \frac{1}{\sqrt{2}}(x_{si} + \frac{\partial}{\partial x_{si}}), \qquad a_{si}^{\dagger} = \frac{1}{\sqrt{2}}(x_{si} - \frac{\partial}{\partial x_{si}}),$$

are oscillator annihilation and creation operators for particles, numbered by $s =$

$1, ..., N$, and motion in the $i = x, y, z$-directions. Thus

$$\sum_{s,i} x_{si} \frac{\partial}{\partial x_{si}} = -\frac{1}{2} \sum_{s,i} a_{si}^{\dagger} a_{si}^{\dagger} + \frac{1}{2} \sum_{s,i} a_{si} a_{si} - \frac{3}{2} N.$$

We are led to consider the algebra generated by the operators

$$A = \frac{1}{2} \sum_{s,i} a_{si}^{\dagger} a_{si}^{\dagger}$$

$$B = \frac{1}{2} \sum_{s,i} a_{si} a_{si},$$

and

$$C = \frac{1}{4} \sum_{s,i} (a_{si}^{\dagger} a_{si} + a_{si} a_{si}^{\dagger}).$$

The A-operator generates the giant monopole mode, (breathing mode) in nuclei. To within one (crucial!) sign these A, B, C, satisfy a commutator algebra similar to the angular momentum algebra. This new algebra is the symplectic algebra sp(2,R), (which is also isomorphic to the su(1,1) and so(2,1) algebras).

The correspondences and differences between the three algebras: the 1-dimensional oscillator (or Heisenberg-Weyl) algebra, associated with translations in phase space, the angular momentum (or su(2)) algebra, associated with rotations in our 3-dimensional world, and the giant monopole, (or sp(2,R)) algebra, associated with dilations or breathing modes are summarized in the table below.

Algebra	1 − dim. oscillator	SU(2)	Sp(2, R)
	$a^\dagger$	J_+	A
Generators	a	J_-	B
	1	J_0	C
Commutation	$[a, a^\dagger] = 1$	$[J_-, J_+] = -2J_0$	$[B, A] = +2C$
relations	$[1, a^\dagger] = 0$	$[J_0, J_+] = +J_+$	$[C, A] = +A$
	$[1, a] = 0$	$[J_0, J_-] = -J_-$	$[C, B] = -B$
Vacua	$\|0\rangle$	$\|-J\rangle$	$\|\Im\rangle$
	$a\|0\rangle = 0$	$J_-\|-J\rangle = 0$	$B\|\Im\rangle = 0$

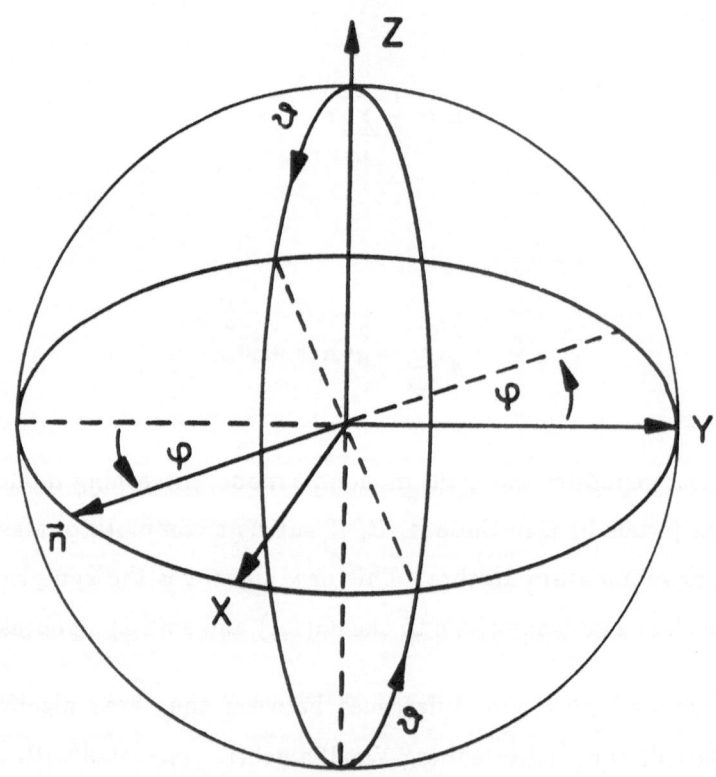

Figure 1.

1. 2 Coherent State Theory for the Angular Momentum Algebra

In this lecture we now want to generalize the simple 1-dimensional harmonic oscillator coherent state to the angular momentum algebra, (the SU(2) algebra), generated by J_+, J_0, J_-. These are not yet the generalized "vector" coherent states which will make up the main subject matter of these lectures; but we shall use the angular momentum coherent state to find the matrix representation of the angular momentum algebra in exactly the same way which will be used later to calculate the matrix representations of higher symmetry algebras. Although we do something quite trivial, ie rederive the angular momentum matrix elements of J_+, J_0, J_-, we will do this in considerable detail to point out various aspects of the coherent state approach. The angular momentum coherent state, (sometimes also called "atomic coherent state" or "Bloch" state; see eg [Arecchi, Courtens, Gilmore, and Thomas, 1972], has wide applications in atomic and solid state physics.

Recall the generalization of the type I coherent state for the angular momentum algebra

$$|\varsigma\rangle = exp(\varsigma^* J_+ - \varsigma J_-)|J, M_{min.} = -J\rangle.$$

In place of the oscillator vacuum, we use the state of minimum M, which is annihilated by J_-; and $a^\dagger \longrightarrow J_+, a \longrightarrow J_-$.

Also, with $\varsigma = \frac{1}{2}\theta e^{i\phi}$,

$$|\varsigma\rangle = e^{-i\theta(\mathbf{J}\cdot\mathbf{n})}|J, -J\rangle,$$

where the rotation operator involves a rotation through the angle θ in a plane normal to the new y-axis, where the new y-axis makes an angle of ϕ with the old one. (Note that the complex number has been named ς to distinguish it from the z to be used presently for a type II coherent state.) Since $\varsigma = \varsigma(\theta, \phi)$, this coherent state can also be labelled by $|\theta, \phi\rangle$. Using the properties of rotation operators we see that

$$|\theta,\phi\rangle = e^{-i\theta(\mathbf{J}\cdot\mathbf{n})}|J,-J\rangle = \sum_M |J,M\rangle D^J_{M,-J}(\phi,\theta,0)^*,$$

where $D^J_{M,M'}$ is the usual rotation matrix. Using the orthonormality of these D-functions, we can see at once that the **1**-operator, to be associated with the $|\theta,\phi\rangle$ is given by

$$\mathbf{1} = \frac{(2J+1)}{4\pi}\int d\Omega|\theta,\phi\rangle\langle\theta,\phi|,$$

where $d\Omega = \sin\theta d\theta d\phi$. Although this "geometrical" form of the coherent state is useful for many atomic calculations, (particularly for thermodynamic quantities), for our purpose, the construction of the matrix representation of the SU(2) algebra, it will be more useful to use the type II coherent state

$$|z\rangle \equiv |z\rangle_{II} = exp(z^* J_+)|-J\rangle.$$

(Note that the relationship between $|z\rangle_I$ and $|z\rangle_{II}$ is now not as trivial as it was for the 1-dimensional oscillator. Another note regarding the notation: We have interchanged the names z and z^* compared with most of the early references, since we want our z to have the same transformation properties as J_+ and the z^* to transform under rotations as J_-; but since z is an arbitrary complex number this is only a question of naming. We will, however, use the name z inside $|z\rangle$).

As for the oscillator, however, we will again map state vectors into z-space functional representations

$$|\psi\rangle \Longrightarrow \psi_J(z) = \langle z|\psi\rangle = \langle -J|e^{zJ_-}|\psi\rangle$$

(note that the value of the quantum number J is now specified in $\psi(z)$).

Operators are mapped into their z-space realizations, $\Gamma(O)$, by

$$O|\psi\rangle \Longrightarrow \Gamma(O)\psi_J(z) = \langle z|O|\psi\rangle = \langle -J|e^{zJ_-}O|\psi\rangle = \langle -J|(e^{zJ_-}Oe^{-zJ_-})e^{zJ_-}|\psi\rangle$$

$$= \langle -J|\left(O + [zJ_-, O] + \frac{1}{2}[zJ_-, [zJ_-, O]] + \cdots\right)e^{zJ_-}|\psi\rangle.$$

The simplest operators are the operators of the algebra itself. 1) For $O = J_-$, only the first term in the commutator expansion survives, and in order to bring down a J_-, we need to differentiate only with respect to z. 2) For $O = J_0$, two terms survive: Action with J_0 to the left on the state of minimum M gives the constant, $-J$, the first commutator gives zJ_-, where we can then use the result of 1): $J_- \to \partial/\partial z$. 3) For $O = J_+$, the first term gives no contribution, since J_+ acting to the left on the minimum M-state gives zero, the first commutator gives $-2zJ_0$ which when acting to the left on the $M = -J$-state gives $2Jz$, while the double commutator, $-z^2 J_-$, now gives a pure z-dependent function, $\longrightarrow -z^2\partial/\partial z$. Thus

$$\Gamma(J_-) = \frac{\partial}{\partial z}$$

$$\Gamma(J_0) = -J + z\frac{\partial}{\partial z}$$

$$\Gamma(J_+) = 2Jz - z^2\frac{\partial}{\partial z}$$

We have thus mapped the angular momentum algebra into an algebra built only from $z, \partial/\partial z$, and constants. Since these satisfy the oscillator commutation relations:

$$[\partial/\partial z, z] = 1, \quad [z, 1] = 0, \quad [\partial/\partial z, 1] = 0,$$

the angular momentum algebra has been mapped into an oscillator algebra. A number of additional remarks may be useful:

1) Note that the $\Gamma(J_i)$ satisfy the angular momentum commutation rules. (This is of course just a check of our arithmetic!)

2) Note that

$$\Gamma(\mathbf{J}^2) = \frac{1}{2}[\Gamma(J_-)\Gamma(J_+) + \Gamma(J_+)\Gamma(J_-)] + \Gamma(J_0)^2 = J(J+1).$$

All z-dependent terms drop out.

3) The above $\Gamma(J_i)$ are by no means unique. This is due to the overcompleteness of the coherent states $|z\rangle$. ($\langle z'|z\rangle \neq \delta$-function). In place of the above z-space realization, we might have used

$$J_+ \longrightarrow z, \quad J_0 \longrightarrow -J + z\frac{\partial}{\partial z}, \quad J_- \longrightarrow 2J\frac{\partial}{\partial z} - z\frac{\partial^2}{\partial z^2}$$

(Note that these satisfy 1) and 2) above). We might also have mapped operators into z-space realizations of the type $\Gamma(z, \partial/\partial z; z', \partial/\partial z')$. These would arise quite naturally via the operator maps

$$\frac{1}{\pi}\int d^2z \, e^{-zz^*}\frac{1}{\pi}\int d^2z' e^{-z'z'^*} |z\rangle\langle z|O|z'\rangle\langle z'|.$$

Such forms may be quite useful in expansions where we want to distinguish the bra and ket sides of the matrix realizations of the operators.

4) This final remark is a crucial one for the later developments: The above z-space realizations, with the simple correspondence $J_- \rightarrow \partial/\partial z$, and the much more complicated correspondence for J_+, is nonunitary. With respect to the Bargmann measure, $\Gamma(J_+)$ is clearly not the hermitian conjugate of $\Gamma(J_-)$; so that the above z-space realization is a nonunitary or Dyson realization of the angular momentum algebra. We can , however, convert the above into a unitary

realization by a similarity transformation with an operator K.

$$\gamma(J_+) = K^{-1}\Gamma(J_+)K$$

$$\gamma(J_-) = K^{-1}\Gamma(J_-)K$$

$$\gamma(J_0) = \Gamma(J_0)$$

Note that, since $\Gamma(J_0)$ was hermitian, no change is needed for J_0. Thus K can be chosen to commute with $\Gamma(J_0)$ and will thus be diagonal in M. Note also that K carries out a very simple role in this example of the angular momentum algebra, with its 1-dimensional ladder of M-values. The operator $\Gamma(J_+)$, increases the power of z, and the M-value, by one unit with each application; but it does not preserve the normalization of the angular momentum eigenstates because of its nonunitary character. Thus K should be the simple command: Reestablish the state normalization. By requiring that the z-space realizations, γ, satisfy

$$\gamma(J_+) = (\gamma(J_-))^\dagger = (K^{-1}\frac{\partial}{\partial z}K)^\dagger = K^\dagger z(K^{-1})^\dagger,$$

and, multiplying from the left with K and from the right with $K^\dagger$,

$$\Gamma(J_+)KK^\dagger = KK^\dagger z.$$

We could now rename the operator $(KK^\dagger)$; but since the simple command: "Multiply the state by a normalization factor", can just as easily be made: "Multiply the state by a **real** normalization factor", we can make K hermitian, $K = K^\dagger$, so that the above simplifies to

$$\Gamma(J_+)K^2 = K^2 z.$$

We could now proceed in either of two ways:

1) By taking matrix elements, (with the Bargmann measure), between the nth and $(n+1)^{th}$ excited states (in the M-ladder, with $M = -J + n$) we could solve this eq. for K^2 directly.

2) A method which will prove more powerful in the generalizations to higher symmetry algebras, makes use of an auxiliary operator, $\Lambda_{op.}$. This is the "Toronto trick". The operator $\Lambda_{op.}$ is to have the property

$$[\Lambda_{op.}, z] = \Gamma(J_+) = 2Jz - z^2 \frac{\partial}{\partial z}.$$

It is easy to solve this commutator relation to get

$$\Lambda_{op.} = 2Jz \frac{\partial}{\partial z} - \frac{1}{2}z^2 \frac{\partial^2}{\partial z^2} = \frac{1}{2}(4J - z\frac{\partial}{\partial z} + 1)(z\frac{\partial}{\partial z}).$$

Even though $\Gamma(J_+)$, acting n times in succession does not give normalized z-space excitations, $z^n/\sqrt{n!}$, it does raise the power of z by one unit each time. Thus $z\partial/\partial z$ has the simple eigenvalue n, leading to the $\Lambda_{eigenvalue}$

$$\Lambda_n = \frac{1}{2}(4J - n + 1)n$$

The equation which determines K^2 thus becomes

$$(\Lambda_{op.}z - z\Lambda_{op.})K^2 = K^2 z.$$

Taking matrix elements, with the Bargmann measure, between the nth normalized z-space function on the right and the $(n+1)^{th}$ on the left, we get

$$(\Lambda_{n+1} - \Lambda_n)\langle \chi_{n+1}|z|\chi_n\rangle K_n^2 = K_{n+1}^2 \langle \chi_{n+1}|z|\chi_n\rangle,$$

which gives

$$\frac{K_{n+1}^2}{K_n^2} = (\Lambda_{n+1} - \Lambda_n) = (2J - n).$$

Starting with $K_0^2 = 1$, since $|-J\rangle$ is assumed normalized, this gives on iteration,

$$K_n^2 = \frac{(2J)!}{(2J-n)!};$$

and, with $\gamma(J_+) = KzK^{-1}$,

$$\langle m|J_+|n\rangle = \langle\chi_m(z)|\gamma(J_+)|\chi_n(z)\rangle = \frac{1}{\pi}\int d^2z \; e^{-zz^*}\frac{z^{*m}}{\sqrt{m!}}(KzK^{-1})\frac{z^n}{\sqrt{n!}}$$

$$= \delta_{m,n+1}\frac{K_{n+1}}{K_n}\langle\chi_{n+1}|z|\chi_n\rangle = \delta_{m,n+1}\sqrt{(2J-n)(n+1)}.$$

Note, in the first step, since we have used the unitary form of the z-space realization of J_+ between orthonormal z-space functions, we can transform to the standard Hilbert space basis. In the second step we have used the value of the K-ratio derived above, and the oscillator matrix element of z to regain this well-known result: With $M = -J + n$ the matrix element of J_+ has its familiar form $\sqrt{(J-M)(J+M+1)}$. Note also that $n_{max.} = 2J$; (with $n = 2J$, the next state has zero norm); therefore also $2J = $ integer. Also

$$\langle\chi_{n-1}|\gamma(J_-)|\chi_n\rangle = \langle\chi_n|\gamma(J_+)|\chi_{n-1}\rangle^* = \sqrt{(2J-n+1)n}.$$

Before examining the angular momentum coherent state in more detail, let us note that we have mapped J_+ into the unitary z-space realization in terms of the K-ratio which was expressed in terms of the quantum number, n, of the initial state

$$J_+ \longrightarrow z\frac{K_{n+1}}{K_n} = z\sqrt{2J-n}.$$

Since n is the eigenvalue of the z-space operator $z\partial/\partial z$, we could also have expressed J_+ solely in terms of z-space operators. Finally, since z and $\partial/\partial z$ are the z-space realizations of oscillator (boson) creation and annihilation operators, J_+ can be expressed in terms of such operators:

$$J_+ \longrightarrow z\sqrt{\left(2J - z\frac{\partial}{\partial z}\right)} \longrightarrow a^\dagger\sqrt{(2J - a^\dagger a)},$$

and its hermitian conjugate

$$J_- \longrightarrow \left(\sqrt{(2J - a^\dagger a)}\right)a.$$

This is the famous Holstein-Primakoff boson realization of the angular momentum algebra. Note, however, the nonuniqueness of such realizations. *Eg,*

$$J_+ \longrightarrow \left(\sqrt{(2J + 1 - a^\dagger a)}\right)a^\dagger; \qquad J_- \longrightarrow a\sqrt{(2J + 1 - a^\dagger a)};$$

would give the same matrix elements and would be the "left square root" counterpart of the above "right square root" form.

From the above matrix elements of J_+ we can now reexpress the coherent state, (recall it is a type II coherent state), by

$$|z\rangle = e^{z^* J_+}|-J\rangle = \sum_{n=0}^{2J} \frac{(z^*)^n}{\sqrt{n!}}\sqrt{\frac{(2J)!}{(2J-n)!}}|-J+n\rangle$$

$$= \sum_{n=0}^{2J} \phi_{J,n}(z)^*|-J+n\rangle, \qquad \text{with} \quad |-J+n\rangle \equiv |J, M = -J + n\rangle.$$

We note at once that the $\phi_{J,n}(z)$ are not normalized with respect to the Bargmann measure. Instead

$$\langle\phi_{J,n}|\phi_{J,m}\rangle = \delta_{nm}\frac{(2J)!}{(2J-n)!} = \delta_{nm}K_n^2.$$

Therefore K_n^2 serves as the normalization integral. To convert the $\phi_{J,n}(z)$ into normalized states, divide by K_n:

$$\chi_n(z) = \frac{z^n}{\sqrt{n!}} = \frac{\phi_{J,n}(z)}{K_n}.$$

In the generalization to problems of higher symmetries, K^2 will become an overlap matrix; and the transformation to an orthonormal basis will involve the square-root taking and inversion of this K^2-matrix.

From the above form of the (type II) coherent state, $|z\rangle$, we can also see that this $|z\rangle$ is not an eigenstate of the step-down operator J_-, since

$$J_-|z\rangle = z^* \sum_{n=0}^{2J} \phi_{J,n}(z)^*(2J - n + 1)|-J + n\rangle$$

$$\neq z^*|z\rangle.$$

The analog of the simple oscillator relation, $a|z\rangle = z^*|z\rangle$ has thus been lost.

Finally, it should be pointed out that the above K-operator method of regaining a unitary realization and an orthonormal basis is not the only solution to the problem of finding the matrix representation of the algebra in question. A second alternative would have been to retain the $\phi_{J,n}(z)$, but change the measure in z-space to enforce the orthonormality of the $\phi_{J,n}$. For the SU(2) algebra it is not only possible to find the new measure; but this measure is also so simple that the necessary integrations can be carried out. For most of the higher symmetry algebras, however, even if the new measure can be found, it is often so complicated that the necessary integrals become very difficult. In general, therefore, it is much simpler to retain the Bargmann measure, with its beautiful Gaussian factor, and use the K-operator or K-matrix method illustrated above.

Before leaving the angular momentum algebra, let us still show that in this case the *new* measure is very simple and show how it can be derived from the type I or angular form of the "atomic" coherent state with which we started this section.

First, let us show that the change of measure

$$\frac{exp(-zz^*)}{\pi} \implies \frac{(2J+1)}{\pi} \frac{1}{(1+zz^*)^{(2J+2)}}$$

leads to the desired orthonormality of the above $\phi_{J,n}$:

With $\qquad z = \varrho e^{i\phi}$

$$\frac{(2J+1)}{\pi} \int d^2z \frac{\phi^*_{J,n}(z)\phi_{J,m}(z)}{(1+zz^*)^{2J+2}} = \delta_{nm}(2J+1)2 \int_0^\infty d\varrho \frac{\varrho^{2n+1}}{(1+\varrho^2)^{2J+2}} \frac{(2J)!}{(2J-n)!n!}$$

$$= \frac{(2J+1)!}{(2J-n)!n!}B(n+1,2J+1-n) = 1,$$

where the Beta function, $B(M,N)$ is defined by

$$B(M,N) = \int_0^\infty \frac{dx x^{M-1}}{(1+x)^{M+N}} = \frac{\Gamma(M)\Gamma(N)}{\Gamma(M+N)}$$

so that the orthonormality follows.

Finally, it can also be shown that the original $\Gamma(J_+) = 2Jz - z^2\partial/\partial z$ is the adjoint of $\Gamma(J_-) = \partial/\partial z$ with respect to this new measure. Let $\psi_a(z)$ and $\psi_b(z)$ be arbitrary analytic functions, given in terms of their expansions

$$\psi_a = \sum_{n=0} a_n z^n; \qquad \psi_b = \sum_{m=0} b_m z^m.$$

Then

$$\frac{(2J+1)}{\pi} \int \frac{d^2z}{(1+zz^*)^{2J+2}}\psi_a^* \frac{\partial}{\partial z}\psi_b = (2J+1)\sum_{n=0}^\infty (n+1)a_n^* b_{n+1} \int_0^\infty \frac{dx x^n}{(1+x)^{2J+2}}$$

$$= \sum_{n=0}^{} a_n^* b_{n+1} \frac{(n+1)!(2J-n)!}{(2J)!};$$

while

$$\frac{(2J+1)}{\pi} \int \frac{d^2 z}{(1+zz^*)^{2J+2}} \psi_b \left[(2Jz - z^2 \frac{\partial}{\partial z}) \psi_a \right]^*$$

$$= (2J+1) \sum_{n=0}^{} (2J-n) b_{n+1} a_n^* \int_0^\infty \frac{dx x^{n+1}}{(1+x)^{2J+2}} = \sum_{n=0}^{} b_{n+1} a_n^* \frac{(2J-n)!(n+1)!}{(2J)!}.$$

To complete our discussion of the angular momentum coherent states, let us give a derivation of the $\frac{(2J+1)}{\pi} \frac{1}{(1+zz^*)^{2J+2}}$ factor for the new measure, and at the same time see the relationship between type I and type II coherent states for this case. The geometrical interpretation of the type I coherent state, with $\varsigma = \frac{\theta}{2} e^{i\phi}$,

$$|\theta, \phi\rangle = e^{\varsigma^* J_+ - \varsigma J_-} |-J\rangle$$

led naturally to the standard angular measure, with

$$1 = \frac{(2J+1)}{4\pi} \int d\Omega |\theta, \phi\rangle \langle \theta, \phi|.$$

To go from this θ, ϕ-dependent type I coherent state to the z-dependent type II coherent state, it is advantageous to make use of the general "disentanglement" relation

$$e^{(a_+ J_+ + a_0 J_0 + a_- J_-)} = e^{b_+ J_+} e^{(lnb_0)J_0} e^{b_- J_-},$$

where for the moment it will be assumed that a_+, a_0, a_-, are arbitrary given complex numbers, and that the b_+, b_0, b_-, are to be determined. For us, actually

$$a_0 = 0, \qquad a_\pm = \pm \frac{\theta}{2} e^{\mp i\phi}.$$

Since the "disentanglement" must follow solely from the angular momentum commutator algebra, and be independent of the J quantum number, ie independent of the irreducible representation label, it is sufficient to use the simplest irreducible representation to carry out the arithmetic. In this case, of course, it means the 2-dimensional irreducible representation, with $J = \frac{1}{2}$

(A side remark: It is surprising that this trick and this angular momentum disentanglement relation are not found in all the books on angular momentum theory, or for that matter on introductory quantum mechanics! - For a reference, undoubtedly not the historically first, see [Gilmore, 1974])

For $J = \frac{1}{2}$:

$$J_+ = \begin{pmatrix} 0 & 1 \\ 0 & 0 \end{pmatrix}; \quad J_- = \begin{pmatrix} 0 & 0 \\ 1 & 0 \end{pmatrix}; \quad J_0 = \begin{pmatrix} \frac{1}{2} & 0 \\ 0 & -\frac{1}{2} \end{pmatrix};$$

so that $(J_+)^n$ and $(J_-)^n$, with $n \geq 2$ are all null matrices.

$$a_+ J_+ + a_0 J_0 + a_- J_- = \mathbf{a} = \begin{pmatrix} \frac{1}{2}a_0 & a_+ \\ a_- & -\frac{1}{2}a_0 \end{pmatrix},$$

and

$$e^{(a_+ J_+ + a_0 J_0 + a_- J_-)} = \mathbf{1}\cosh a + \frac{\mathbf{a}}{a}\sinh a,$$

or

$$e^{(a_+ J_+ + a_0 J_0 + a_- J_-)} =$$

$$\begin{pmatrix} \cosh a + \frac{1}{2}\frac{a_0}{a}\sinh a & \frac{a_+}{a}\sinh a \\ \frac{a_-}{a}\sinh a & \cosh a - \frac{1}{2}\frac{a_0}{a}\sinh a \end{pmatrix}$$

$$= \sqrt{\frac{1}{b_0}} \begin{pmatrix} b_0 + b_- b_+ & b_+ \\ b_- & 1 \end{pmatrix}.$$

In the above, the quantity a is defined by

$$a^2 \equiv \frac{1}{4}a_0^2 + a_+ a_-,$$

but in our special case,

$$a_0 = 0; \qquad a_\pm = \pm\frac{\theta}{2}e^{\mp i\phi}; \qquad a = i\frac{\theta}{2}.$$

Comparison of the matrix elements gives the solution

$$b_\pm = \pm\tan\frac{\theta}{2}e^{\mp i\phi}$$

$$b_0 = \frac{1}{\cos^2\frac{\theta}{2}} = 1 + \tan^2\frac{\theta}{2}$$

In the final disentangled form b_+ is to be identified with z^*. Thus

$$\tan^2\frac{\theta}{2} = zz^*$$

and

$$e^{b_+ J_+}e^{(ln b_0)J_0}e^{b_- J_-}|-J\rangle$$

$$= e^{z^* J_+}e^{(ln(1+zz^*))(-J)}1|-J\rangle = e^{z^* J_+}\frac{1}{(1+zz^*)^J}|-J\rangle$$

In addition

$$\sin\theta d\theta d\phi = 4\varrho d\varrho d\phi\frac{1}{(1+\varrho^2)^2},$$

where

$$z = \varrho e^{i\phi}, \qquad \text{so that} \qquad \varrho = \tan\frac{\theta}{2}.$$

Thus

$$\frac{(2J+1)}{4\pi}\int d\Omega|\theta,\phi\rangle\langle\theta,\phi|$$

$$= \frac{(2J+1)}{\pi}\int_0^{2\pi}d\phi\int_0^\infty\frac{\varrho d\varrho}{(1+\varrho^2)^2}\frac{e^{z^* J_+}}{(1+\varrho^2)^J}|-J\rangle\langle-J|\frac{e^{zJ_-}}{(1+\varrho^2)^J}$$

$$= \frac{(2J+1)}{\pi}\int\frac{d^2z}{(1+zz^*)^{2J+2}}|z\rangle\langle z|,$$

which is the "atomic" z-space measure which we wanted to substantiate.

1. 3   The Giant Monopole Algebra, $Sp(2, R)$.

The dilation operation, and its associated spherically symmetric breathing modes or giant monopole excitations, led us to the $Sp(2, R)$ symmetry algebra. This algebra is generated by the operators A, B, C, of section 1.1. Its commutator algebra differs from the angular momentum algebra in only one sign. But this change of sign is crucial for the eigenvalue spectrum and the physics associated with this symmetry. With $B = A^\dagger$, and

$$[B, A] = +2C; \qquad [C, A] = +A; \qquad [C, B] = -B;$$

the operators again can be organized into a raising, a lowering, and a "weight" operator. We will assume that there is a "lowest" state which is annihilated by the lowering operator B and is specified by the quantum number $\mathfrak{S}$.

$$B|\mathfrak{S}\rangle = 0; \qquad C|\mathfrak{S}\rangle = \mathfrak{S}|\mathfrak{S}\rangle.$$

For the realization of this algebra in terms of oscillator creation and annihilation operators for an N-particle system in three dimensions, given in section 1.1, the existence of a "lowest" state is apparent. Any nuclear shell model valence state, (with no core excitations), is annihilated by the B-operator which decreases the total number of oscillator quanta by 2 units. It is also clear that the quantum number $\mathfrak{S}$ is not necessarily an integer or half-integer, since

$$C = \frac{1}{2} \sum_{s,i} a_{si}^\dagger a_{si} + \frac{3N}{4} \quad = \quad \frac{1}{2} \mathbf{N}^{osc.quanta} + \frac{3N}{4},$$

where $\mathbf{N}^{osc.quanta}$ counts the total number of oscillator quanta of the system. (Actually, if we want to exclude the motion of the center of mass, the s-sum over N single particle coordinates can be replaced by a sum over $N - 1$ *relative* motion Jacobi coordinates, so that N in the above should be replaced by $N - 1$. However, the eigenvalues of C can still be $\frac{1}{4}$-integer.)

As for the angular momentum algebra, we define a type II coherent state through the raising operator, A, by

$$|z\rangle = e^{z^*A}|\Im\rangle.$$

Again, state vectors $|\psi\rangle$ and operators O are mapped into their z-space realizations by

$$|\psi\rangle \Longrightarrow \langle z|\psi\rangle = \psi_\Im(z) = \langle\Im|e^{zB}|\psi\rangle$$

$$O|\psi\rangle \Longrightarrow \Gamma(O)\psi_\Im(z) = \langle\Im|\left(O + [zB, O] + \frac{1}{2}[zB, [zB, O]] + \cdots\right)e^{zB}|\psi\rangle.$$

The operators of the algebra itself again require only the first few terms of this commutator expansion. With the hermitian operator, C, yielding the eigenvalue $\Im$ when acting to the left on the lowest or starting state, we get the z-space realizations

$$\Gamma(B) = \frac{\partial}{\partial z}$$

$$\Gamma(C) = \Im + z\frac{\partial}{\partial z}$$

$$\Gamma(A) = 2\Im z + z^2\frac{\partial}{\partial z}$$

From this z-space realization we can see at once that the quadratic invariant operator, (Casimir operator), is now

$$\Gamma(C)^2 - \frac{1}{2}(\Gamma(A)\Gamma(B) + \Gamma(B)\Gamma(A)) = \Im(\Im - 1).$$

The above is again a nonunitary or Dyson realization of the algebra. (Note that we want to retain the Bargmann measure and therefore define the adjoint operation with respect to this measure). To convert the above to a unitary (or Holstein-Primakoff) realization, we again use a K-operator to convert the nonunitary $\Gamma(O)$ into the unitary $\gamma(O)$

$$\gamma(A) = K^{-1}\Gamma(A)K; \qquad \gamma(B) = K^{-1}\Gamma(B)K.$$

Since $\Gamma(C)$ is self-adjoint K can again be made to commute with $\Gamma(C)$. As before, we will try to perform the unitarization with a *hermitian* K, *ie* $K^\dagger = K$. The requirement $\gamma(A) = (\gamma(B))^\dagger$ leads to

$$\gamma(A) = \left(K^{-1}\frac{\partial}{\partial z}K\right)^\dagger = KzK^{-1} = K^{-1}\Gamma(A)K.$$

Multiplying by K from the left and by $K^\dagger = K$ from the right, this leads to

$$\Gamma(A)K^2 = K^2 z.$$

We again solve this equation for K^2 with the use of the auxiliary operator $\mathbf{\Lambda}_{op.}$ with the property

$$[\mathbf{\Lambda}_{op.}, z] = \Gamma(A) = 2\Im z + z^2 \frac{\partial}{\partial z}.$$

This operator equation has the simple solution

$$\mathbf{\Lambda}_{op.} = 2\Im z \frac{\partial}{\partial z} + \frac{1}{2}z^2 \frac{\partial^2}{\partial z^2} = \frac{1}{2}(4\Im + z\frac{\partial}{\partial z} - 1)(z\frac{\partial}{\partial z}).$$

As for the angular momentum algebra, every action with $\Gamma(A)$ raises the power of z by one. The spectrum of C-eigenvalues beginning with $\Im$ will have reached $\Im + n$ after n such raising operations, so that the eigenvalue of $z\frac{\partial}{\partial z}$ is again simply

n. This leads to the eigenvalue

$$\Lambda_n = \frac{1}{2}(4\Im + n - 1)n,$$

with

$$\Lambda_{n+1} - \Lambda_n = (2\Im + n).$$

The equation which determines K^2,

$$(\Lambda_{op.}z - z\Lambda_{op.})K^2 = K^2 z,$$

again leads to the matrix element equation

$$(\Lambda_{n+1} - \Lambda_n)\langle\chi_{n+1}|z|\chi_n\rangle K_n^2 = K_{n+1}^2\langle\chi_{n+1}|z|\chi_n\rangle.$$

Now this gives

$$\frac{K_{n+1}^2}{K_n^2} = (2\Im + n),$$

which differs from the $SU(2)$ result by the $+$ sign in front of the n. Now

$$\langle\chi_m(z)|\gamma(A)|\chi_n(z)\rangle = \langle\chi_m|KzK^{-1}|\chi_n\rangle = \delta_{m,n+1}\frac{K_{n+1}}{K_n}\langle\chi_{n+1}|z|\chi_n\rangle,$$

or, transforming from this unitary form of A between orthonormal z-space functions, to the standard Hilbert space basis,

$$\langle\Im, m|A|\Im, n\rangle = \delta_{m,n+1}\sqrt{(2\Im + n)(n + 1)}.$$

Note that now n can grow indefinitely. There is no upper value of n for which K_{n+1} becomes zero. The spectrum of C-eigenvalues is given by $\Im + n$, with

$n = 0, 1, \ldots, \to \infty$. The coherent state can be expressed by

$$|z\rangle = e^{z^* A}|\Im\rangle = \sum_{n=0}^{\infty} \frac{(z^*)^n}{\sqrt{n!}} \sqrt{2\Im(2\Im + 1) \cdots (2\Im + n - 1)} |\Im, n\rangle$$

$$= \sum_{n=0}^{\infty} \phi_{\Im,n}^*(z)|\Im, n\rangle,$$

where

$$\phi_{\Im,n}(z) = \frac{z^n}{\sqrt{n!}} \sqrt{\frac{\Gamma(2\Im + n)}{\Gamma(2\Im)}}.$$

These $\phi_{\Im,n}$ are now orthonormal with respect to a complicated z-space measure involving a modified Bessel function of the third kind, see [Barut and Girardello, 1971]. In actual practice it is much simpler to work with a scalar product with the standard Bargmann measure, and use the K^2-method to gain an orthonormal basis. The orthonormal z-space functions are again given by

$$\frac{\phi_{\Im,n}(z)}{K_n} \qquad \text{where} \qquad K_n = \sqrt{\frac{\Gamma(2\Im + n)}{\Gamma(2\Im)}}.$$

Note that we have used the Γ-function since $2\Im$ is not necessarily an integer. Note also that

$$B|z\rangle \neq z^*|z\rangle.$$

On the other hand, a type III coherent state, with $B|z\rangle_{III} = z^*|z\rangle_{III}$ could have been constructed through

$$|z\rangle_{III} = \sum_{n=0}^{\infty} \frac{z^{*n}}{\sqrt{n!}} \sqrt{\frac{\Gamma(2\Im)}{\Gamma(2\Im + n)}} |\Im, n\rangle.$$

2. The Vector Coherent State Method.

Generalization to Higher Symmetries.

The 1-dimensional oscillator algebra, the angular momentum algebra, and the giant monopole algebra of section 1 were all 1-dimensional in the sense that the ladders of eigenvalues (weight spaces) were all 1-dimensional ladders. The algebras all had a single raising, a single lowering, and a single weight (J_0-like) operator. For the higher symmetry algebras we will now generalize this and try to organize these algebras into a family of raising operators, their hermitian conjugate lowering operators, and a family which is the generalization of the J_0 operator. The latter are to form a subalgebra of the full algebra, (subgroup of the full group).

$$
\begin{array}{lllr}
J_+ & \Longrightarrow & A_1, A_2, ..., A_m & \text{Raising Ops.} \\
J_- & \Longrightarrow & B_1, B_2, ..., B_m & \text{Lowering Ops.} \\
J_0 & \Longrightarrow & \mathbf{C} & \text{Core subgroup Ops.}
\end{array}
$$

It will further be assumed that these operators satisfy the following properties:

1) $\qquad\qquad B_i = (A_i)^\dagger$

2) $\qquad\qquad [A_i, A_j] = 0 \qquad\qquad$ for all i, j

3) $\qquad\qquad \mathbf{C} = \text{Subalgebra}$
   (containing Cartan subalgebra of full algebra)

4) $\qquad\qquad B_i|(\varrho)\eta\rangle = 0 \qquad\qquad$ for all i, all η

Condition 4) states that the B_i are "vacuum annihilation operators" for the "vacua", $|(\varrho)\eta\rangle$, where (ϱ) denotes the quantum numbers which specify an irreducible representation of $\mathbf{C}$, and η the additional quantum numbers needed to specify the states of (ϱ). *Eg*, if $\mathbf{C} = SU(2)$, then (ϱ) would be a single angular momentum quantum number of type J, while η would be an M and would run over the $(2J+1)$ values $M = J \rightarrow -J$. In general, η takes on values η_k, with

$k = 1, 2, ..., dimension(\varrho) \equiv dim(\varrho)$. The "vacua" thus become "vectors", or $dim(\varrho)$-fold degenerate vacua, and the coherent state

$$|\mathbf{z}\rangle = exp(\sum_{i=1}^{m} z_i^* A_i)|(\varrho)\eta\rangle,$$

where $|\mathbf{z}\rangle \equiv |z_1, ..., z_m; (\varrho)\eta\rangle$, has also become a $dim(\varrho)$-dimensional vector.

In the above it will further be assumed that the core subalgebra, C, contains the Cartan subalgebra of the full algebra, *ie* the family of hermitian commuting operators which lead to additive M-type quantum numbers within C are the same as (or linear combinations of) such operators within the full algebra. Although some of the above restrictions, especially restriction 2), the commutability of the A_i, can be relaxed, - (much progress along these lines has been made very recently, see [Rowe et al.; 1987 preprint] and [Le Blanc and Rowe; 1987 preprints a,b])-, these lectures will discuss only the simpler cases satisfying the above conditions. Despite these restrictions, we shall see that there are a large number of interesting algebras of the above type; and with the lifting of some of the restrictions almost all of the classical Lie algebras can be treated by the vector coherent state technique.

A word about the language: Although we have named the A_i "raising" operators and the B_i "lowering" operators to preserve the language of the 1-dimensional algebras, this language may be confusing. What is "up" or "down" is often a question of definition, depending on signs within the defining operators. It may therefore be better to name the A_i "creation"-type operators, and the B_i "annihilation"-type operators. The B_i are always the operators which annihilate the "vacua". This also gives us the freedom of constructing our states, starting from the "tops" *or* from the "bottoms", *or* from any extremal corners of the weight space.

We shall start by giving a few examples of algebras which satisfy the criteria 1) to 4):

The $Sp(6, R)$ Algebra.

An obvious example of an algebra of the above type is a generalization of the giant monopole algebra, $Sp(2, R)$. This algebra can be made to include both monopole and quadrupole excitations, if we do not sum over $i = x, y, z$ to construct a spherically symmetric, $L = 0$ $2\hbar\omega$ excitation operator, A ; but instead, for fixed i, j, consider

$$A_{ij} = \sum_{s=1}^{N} a_{si}^{\dagger} a_{sj}^{\dagger} \qquad = A_{ji}$$

$$B_{ij} = \sum_{s=1}^{N} a_{si} a_{sj} \qquad = B_{ji}$$

$$C_{ij} = \frac{1}{2} \sum_{s=1}^{N} (a_{si}^{\dagger} a_{sj} + a_{sj} a_{si}^{\dagger})$$

where the $a_{si}^{\dagger}$, a_{si} are again oscillator creation and annihilation operators for particles labelled by particle number $s = 1, ..., N$ and motion in the $i = x, y, z$-direction. The 9 C_{ij} form the subgroup $U(3)$. Together with the 6 A_{ij} and the 6 B_{ij}, these 21 operators generate the full group $Sp(6, R)$. This is the noncompact version of the unitary symplectic group $Sp(6)$. Similar to the relationship between $Sp(2, R)$ and $SU(2)$, the generators of $Sp(6, R)$ and $Sp(6)$ satisfy the same commutator algebra to within a few crucial changes in sign. The 21 generators of $Sp(6, R)$ can also be expressed as linear combinations of the operators

$$\sum_{s=1}^{N} x_{si} x_{sj} \qquad \longrightarrow \qquad \left(\sum_{s} r_s^2 \ ; \ \sum_{s} r_s^2 Y_{2\mu}(\theta_s, \phi_s) \right)$$

the corresponding

$$\sum_{s=1}^{N} p_{si} p_{sj}$$

as well as

$$\sum_{s=1}^{N}(x_{si}p_{sj} + x_{sj}p_{si})$$

and

$$\mathbf{L} = \sum_{s=1}^{N}(x_{si}p_{sj} - x_{sj}p_{si})$$

Note that this algebra contains the N-particle monopole operator, $\sum r_s^2$, the quadrupole operators, the analagous momentum combinations, and the orbital angular momentum operator. The $U(3)$ subgroup, generated by $\frac{1}{2}\sum_s(x_{si}x_{sj} + p_{si}p_{sj})$ and $\mathbf{L}$, is the Elliott $U(3)$ group.

The Unitary Groups $U(n) \supset U(n-1) \times U(1)$.

Other examples of symmetries which can be classified according to the above scheme, and fit the requirements 1)-4), include the unitary groups $U(n)$, with subgroups $\mathbf{C} = U(n-1) \times U(1)$. The generators of $U(n)$ can be constructed either from boson operators $a_{si}^\dagger$, a_{si}, as in the above C_{ij} of the $U(3)$ group; or, alternately, in other applications may follow naturally from constructions in terms of fermion creation and annihilation operators. The quantum numbers specifying the fermion creation and annihilation operators $b_{\alpha k}^\dagger$, $b_{\alpha k}$ often divide themselves naturallly into two sets, α and k. Eg, α might stand for the isospin quantum numbers, $\alpha \equiv m_t = \pm\frac{1}{2}$, while k stand for the space-spin quantum numbers of a nucleon in the j-shell of the nuclear shell model, so that $k \equiv m_j$. Then

$$E_{ik} = \sum_{\alpha} b_{\alpha i}^\dagger b_{\alpha k} \qquad\qquad i, k = 1, ..., (2j+1)$$

are the generators of $U(2j+1)$. On the other hand,

$$E_{\alpha\beta} = \sum_{k} b_{\alpha k}^\dagger b_{\beta k} \qquad\qquad \alpha, \beta = m_t = \pm\frac{1}{2}$$

are the generators of $U(2)$. Such E_{ik}, or $E_{\alpha\beta}$, which annihilate a particle in state k and replace it with one in state i are already written in Cartan standard notation. With $i < k$ the E_{ik} are shift operators, corresponding to positive roots $(e_i - e_k)$, and the $E_{ki} = (E_{ik})^\dagger$ are shift operators, corresponding to negative roots $-(e_i - e_k)$, while the $E_{ii} \equiv H_i$ form the Cartan subalgebra of commuting, hermitian operators. (Note again, however, that the words "positive" and "negative" could just as well be applied to the case $i > k$ and like the words "raising" and "lowering" are ambiguous). The E_{ij} satisfy the $U(n)$ commutator algebra

$$[E_{ij}, E_{kl}] = \delta_{jk} E_{il} - \delta_{il} E_{kj}.$$

From these commutation relations it can be seen that the E_{ik} can be organized into:

E_{ni}	$i = 1, ..., n-1;$	A_i	"Creation" Ops.
E_{in}	$i = 1, ..., n-1;$	B_i	"Annihilation" Ops.
$E_{ij}, \quad E_{nn}$	$i, j \leq n-1;$	C	$U(n-1) + U(1)$ Gens.

Alternately, with E_{ik} replaced by traceless E_{ik}, a similar classification follows for $SU(n)$, where the core subalgebra C is replaced by $SU(n-1) + U(1)$, generated by traceless E_{ij}, with $i, j \leq n-1$, and $\frac{1}{n}[\sum_i^{n-1} E_{ii} - (n-1)E_{nn}]$. In either case, the requirements 1) to 4) are satisfied for this very general group category.

The Orthogonal Groups $SO(n) \supset SO(n-2) \times SO(2)$.

Another very general category is given by the rotation groups in n dimensions, $SO(n)$, generated by n-dimensional angular momentum operators

$$J_{jk} = -i\left(x_j \frac{\partial}{\partial x_k} - x_k \frac{\partial}{\partial x_j}\right) = -J_{kj}, \qquad with \quad j, k = 1, ..., n$$

where these are hermitian, $J_{jk} = (J_{jk})^\dagger$. The J_{jk} satisfy the commutator algebra

$$[J_{jk}, J_{pq}] = i(\delta_{jp} J_{kq} + \delta_{kq} J_{jp} - \delta_{jq} J_{kp} - \delta_{kp} J_{jq})$$

From these commutation relations it can again be seen that these algebras also can be organized according to our scheme:

$$(J_{j(n-1)} + iJ_{jn}) \qquad j = 1,...,n-2 \qquad A_j \qquad \text{"Creation" Ops.}$$
$$(J_{j(n-1)} - iJ_{jn}) \qquad j = 1,...,n-2 \qquad B_j \qquad \text{"Annihilation" Ops.}$$
$$J_{jk}, \quad J_{(n-1)n} \qquad j,k = 1,...,n-2 \qquad C \qquad SO(n-2) + SO(2) \text{ Gens.}$$

Although these are somewhat abstract, we shall see that the specific cases, $n = 5$, $n = 8$, and $n = 7$ have interesting applications in nuclear physics through the neutron-proton quasispin algebra, the LST- pairing algebra, and the $SO(8) \supset SO(6)$ and $SO(7) \supset SO(5)$ branches of the Ginocchio S,D fermion pair algebra.

Before searching for more very general but abstract examples, let us study some very specific examples in considerable detail to see how the standard coherent state theory is generalized to give very explicit matrix representations of these higher symmetries described by vector coherent states.

3. Detailed Examples

3.1 The Group $SU(3)$.

As our first very specific example let us consider the group $SU(3)$. Two of the historically most important realizations of this symmetry are the Elliott $SU(3)$ model and the $SU(3)$-flavor quark model, (an outgrowth of the Sakata model).

In our scheme the 8 generators of $SU(3)$ will be organized into two raising or creation-type operators, two lowering or annihilation-type operators, and the generators of the core subgroup, $SU(2)+U(1)$:

A_i : E_{31} E_{32}

B_i : E_{13} E_{23}

C : E_{12}, E_{21}, $\frac{1}{2}(E_{11} - E_{22})$; $\frac{1}{3}(E_{11} + E_{22} - 2E_{33})$

In the $SU(3)$-flavor quark model, the generators

$$E_{ik} = \sum_\alpha b_{\alpha i}^\dagger b_{\alpha k} \qquad \text{with} \quad \alpha = s\, m_s, \ell\, m_\ell, color$$

involve a sum over the full set of spin, space, color quantum numbers; and the quantum numbers i, k signify: $i = 1$ for *up*, $i = 2$ for *down*, and $i = 3$ for *strange* quark. Thus E_{31} annihilates an up quark and replaces it with a strange quark with the same spin, space, and color quantum numbers. The E_{ii} count the number of quarks of type i. The generators of the $SU(2)$ subgroup which can convert up into down quarks and vice versa are now the components of the isospin vector **I**; while the $U(1)$ generator is the hypercharge, Y, which essentially counts the number of strange quarks.

$$E_{12} = I_+, \quad \frac{1}{2}(E_{11} - E_{22}) = I_0, \quad E_{21} = I_-; \qquad \frac{1}{3}(E_{11} + E_{22} - 2E_{33}) = Y.$$

(Note that we have called E_{31} a "raising" or "creation" type operator. By creating a strange quark it does indeed raise the energy; but note that it lowers the value of the Y quantum number by one unit).

In the Elliott $SU(3)$ model, with the generators expressed in terms of oscillator creation and annihilation operators, $E_{ik} = \frac{1}{2}\sum_s(a_{si}^\dagger a_{sk} + a_{sk}a_{si}^\dagger)$, the i,k now stand for the three space directions in the body-fixed (primed) frame of a deformed nucleus: $i = x', y', z'$; for $i = 1,2,3$. Now the physically interesting quantum number is $\epsilon = (2E_{33} - E_{11} - E_{22}) = 2N_{z'} - N_{x'} - N_{y'}$ which is related to the strength of the intrinsic quadrupole moment in the deformed system. Note that $N_{z'}$ counts the number of oscillator quanta for motion in the z' direction. Now E_{31} shifts oscillator quanta from the x' direction into the z' direction and raises the value of the intrinsic quadrupole moment. Note that $\epsilon \leftrightarrow -3Y$. In the Elliott intrinsic nuclear frame, E_{12}, $\frac{1}{2}(E_{11} - E_{22})$, and E_{21} are the components of the so-called Λ angular momentum which classifies the symmetry of the motions in the intrinsic x', y' plane.

In both realizations of the $SU(3)$ model we will choose the "vacuum" states $|(\varrho)\eta\rangle$, the "lowest" or starting states, to be states with eigenvalue $E_{33} = 0$; ie states with up and down quarks only in the quark model, or states with only x' and y' oscillator excitations in the Elliott model, (outside of closed shells which have equal numbers of z', x', y' oscillator quanta). These "lowest" states have a permutation symmetry describable by a two-rowed Young tableau, with $\lambda + \mu$ squares in the first row and μ squares in the second row. (We use the Elliott notation; see Fig.2 below)

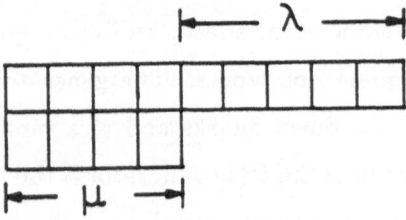

Figure 2.

In the case of the quark model this tableau describes the symmetry of the flavor part of the wave function of a system of $(\lambda + 2\mu)$ up and down quarks. In the case of the $SU(3)$ shell model it describes the symmetry of the $(\lambda + 2\mu)$ oscillator quanta carried by particles outside of inner closed shells. Since they are

x', y' quanta only this would be an oblate configuration of a nucleus. Recall that squares in the same column of a Young tableau imply possible antisymmetrization. (In the case of the quark model this would mean antisymmetrization of a pair of up, down quarks). Squares in the same row of the tableau, on the other hand, imply possible symmetrization. In the "lowest" state of up, down quarks only, we thus have μ pairs coupled to $I = 0$, and an additional number of λ quarks which could be all up, all down, or any combination of up and down, and therefore have M_I-values ranging from $+\frac{1}{2}\lambda \rightarrow -\frac{1}{2}\lambda$. They are therefore the members of an $I = \frac{1}{2}\lambda$ isospin multiplet. The "vacuum" states $|(\varrho)\eta\rangle$ are therefore specified by the quantum numbers

$$Y_{vac.} = \frac{1}{3}(\lambda + 2\mu), \quad I_{vac.} = \frac{1}{2}\lambda, \quad M_I; \quad \text{or :} \quad |(\varrho)\eta\rangle = \left|(\lambda,\mu), I = \frac{\lambda}{2}, M_I\right\rangle,$$

with

$$E_{i3}\left|(\lambda,\mu), I = \frac{\lambda}{2}, M_I\right\rangle = 0 \quad for \quad i = 1,2; \quad M_I = \frac{1}{2}\lambda \rightarrow -\frac{1}{2}\lambda.$$

The vector coherent state is now constructed in terms of the two complex variables, z_1, z_2,

$$|z\rangle = e^{(z_1^* E_{31} + z_2^* E_{32})}|(\lambda,\mu), M_I\rangle.$$

State vectors are to be mapped into z-space functions

$$|\psi\rangle \Longrightarrow \psi_{(\lambda,\mu)M_I}(z) = \langle z|\psi\rangle = \langle(\lambda,\mu)M_I|e^{(z_1 E_{13} + z_2 E_{23})}|\psi\rangle.$$

It will be convenient to introduce the shorthand notation

$$X \equiv (z_1 E_{13} + z_2 E_{23}).$$

Operators are mapped into their z-space realizations, $\Gamma(O)$, via

$$O|\psi\rangle \Longrightarrow \Gamma(O)\psi(z) = \langle(\lambda,\mu)M_I|e^X O|\psi\rangle = \langle(\lambda,\mu)M_I|(e^X O e^{-X})e^X|\psi\rangle$$

$$= \langle(\lambda,\mu)M_I|(O + [X,O] + \frac{1}{2}[X,[X,O]] + \cdots)e^X|\psi\rangle.$$

The most interesting operators are again the 8 generators of the $SU(3)$ algebra itself. The annihilation-type operators, E_{13} and E_{23}, commute with $X = (z_1 E_{13} + z_2 E_{23})$. For these operators only the first terms in the commutator expansion survive and therefore again lead to the very simple z-space realizations, involving only the partial derivatives. We shall use the shorthand notation $\partial/\partial z_i \equiv \partial_i$. For the generators of the subalgebra, $SU(2) + U(1)$, only the first two terms in the commutator expansion survive. Now, however, only the members of the Cartan subalgebra, the operators $\frac{1}{2}(E_{11} - E_{22})$ and $\frac{1}{3}(E_{11} + E_{22} - 2E_{33})$, can be replaced by quantum numbers of the "vacuum" or starting states; while the operators, $E_{12} = I_+$ and $E_{21} = I_-$, effect changes in the M_I. However, they do not change the quantum numbers of $Y_{vac.} = \frac{1}{3}(\lambda + 2\mu)$ or $I_{vac.} = \frac{1}{2}\lambda$. They can change only the orientation of the isospin of the vacuum iso-vector. They act like the components of an intrinsic spin operator. The isovector $\mathbf{I}$ operators acting to the left on the components of the vacuum multiplet will therefore be named the "intrinsic" isospin operators, to be denoted by an $\mathbf{\it I}$. A similar notation, $\mathbf{\it Y}$, or $\mathbf{\it E}_{ik}$ for any component of the core subgroup, will be used when these act only on the vacuum states. The commutator expansion in terms of X thus yields

$$\Gamma(E_{13}) = \partial_1 \quad ; \qquad \Gamma(E_{23}) = \partial_2$$

$$\left. \begin{aligned} \Gamma(E_{12}) &= I_+ - z_2 \partial_1 \\ \Gamma(\tfrac{1}{2}(E_{11} - E_{22})) &= I_0 - \tfrac{1}{2}(z_1 \partial_1 - z_2 \partial_2) \\ \Gamma(E_{21}) &= I_- - z_1 \partial_2 \end{aligned} \right\} = \mathbf{\it I}^{intr.} + \mathbf{I}^{coll.}$$

$$\Gamma(\frac{1}{3}(E_{11} + E_{22} - 2E_{33})) = \mathbf{\it Y} - (z_1 \partial_1 + z_2 \partial_2) \qquad = \mathbf{\it Y}^{intr.} + Y^{coll.}$$

$$\Gamma(E_{31}) = z_1 (I_0 + \frac{3}{2}Y) + z_2 I_- - z_1^2 \partial_1 - z_1 z_2 \partial_2$$

$$\Gamma(E_{32}) = z_1 I_+ + z_2 (-I_0 + \frac{3}{2}Y) - z_2 z_1 \partial_1 - z_2^2 \partial_2,$$

where the creation type operators, $\Gamma(E_{3i})$, with $i = 1, 2$ could also be written

$$\Gamma(E_{3i}) = z_\alpha E_{\alpha i} - z_i E_{33} - z_i z_\alpha \partial_\alpha.$$

Now, repeated indices α imply a sum over indices $1, 2$.

With this z-space realization we have mapped the $SU(3)$ algebra onto a 2-dimensional oscillator algebra, generated by $z_1, \partial_1; z_2, \partial_2;$ with the additional operators Y, I_k. These "intrinsic" operators act only on the vacuum starting states and hence commute with the z-space operators

$$[Y, z_i] = 0; \qquad [Y, \partial_i] = 0; \qquad [I_k, z_i] = 0; \qquad [I_k, \partial_i] = 0.$$

The $SU(3)$ algebra has thus been mapped into a direct sum of a 2-dimensional oscillator algebra and an intrinsic $U(1) + SU(2)$ algebra. Note that the operators I are composed of an intrinsic part and a purely z-dependent part

$$I_+^{coll.} = -z_2 \partial_1; \qquad I_0^{coll.} = -\frac{1}{2}(z_1 \partial_1 - z_2 \partial_2); \qquad I_-^{coll.} = -z_1 \partial_2.$$

Although the names "collective" and "intrinsic" have their origin in the $Sp(6, R)$ symmetry, which was the first to be treated by vector coherent state techniques, they are appropriate for many of the other applications. *Eg* here, in the $SU(3)$-flavor quark model, the intrinsic isospin operator changes only the orientation of the isospin vector of the state made up only of up and down quarks, since the intrinsic isospin operator acts only on the state where all quarks are of the up and down variety. The "collective" part of the isospin operator, however, changes the isospin by converting up or down i-spin $\frac{1}{2}$ quarks into i-spin 0 strange quarks. Such an operation thus involves the family of quarks collectively. The operator

$$Y^{coll.} = -(z_1 \partial_1 + z_2 \partial_2)$$

counts the number of strange quarks. The orthonormal Bargmann space functions are the 2-dimensional oscillator functions

$$Z(z) = Z(z_1, z_2) = \frac{z_1^a}{\sqrt{a!}} \frac{z_2^b}{\sqrt{b!}} = \frac{z_1^{\frac{1}{2}w-m}}{\sqrt{(\frac{1}{2}w - m)!}} \frac{z_2^{\frac{1}{2}w+m}}{\sqrt{(\frac{1}{2}w + m)!}},$$

where the meaning of the labels a and b follows from

$$(z_1\partial_1 + z_2\partial_2)Z(z) = wZ(z) = (a + b)Z(z),$$

$$I_0^{coll.} Z(z) = -\frac{1}{2}(z_1\partial_1 - z_2\partial_2)Z(z) = mZ(z) = -\frac{1}{2}(a - b)Z(z).$$

The degree, w, of the z-space function counts the number of strange quarks. The hypercharge, Y, is decreased by 1 unit every time an up or down quark is converted to a strange one. The isospin character of the z-space function follows from the well known Schwinger-Bargmann construction of the angular momentum eigenfunctions. (Note, however, that z_1, z_2 have interchanged their conventional positions). Direct application also shows that

$$\left(\frac{1}{2}(I_+^{coll.} I_-^{coll.} + I_-^{coll.} I_+^{coll.}) + (I_0^{coll.})^2\right) Z_m^{\frac{1}{2}w}(z) = \frac{w}{2}(\frac{w}{2} + 1)Z_m^{\frac{1}{2}w}(z).$$

The $Z(z)$ are thus I-spin eigenfunctions with $I^{coll.} = \frac{1}{2}w$. From their explicit form it can be seen that the angular momentum-reduced matrix element of z has the simple oscillator value $\langle \frac{1}{2}(w + 1)\|z\|\frac{1}{2}w \rangle = \sqrt{(w + 1)}$. By choosing $m_{initial} = \frac{1}{2}w$:

$$\frac{1}{\pi} \int d^2 z_2 e^{-z_2 z_2^*} \frac{(z_2^*)^{w+1}}{\sqrt{(w + 1)!}} z_2 \frac{z_2^w}{\sqrt{w!}} = \sqrt{(w + 1)}$$

$$= \left\langle \frac{1}{2}(w + 1)\left\|z\right\|\frac{1}{2}w \right\rangle \left\langle \frac{w}{2} \frac{w}{2} \frac{1}{2}\frac{1}{2}\left|\frac{1}{2}(w + 1)\frac{1}{2}(w + 1)\right. \right\rangle,$$

where the m-values have been chosen so that the Wigner coefficient has the value $+1$.

The full orthonormal z-space functions, $\chi(z)$, will be constructed by angular momentum coupling

$$\left[Z^{\frac{w}{2}}(z)\left|\frac{\lambda}{2}\right\rangle\right]^I_{M_I} = \sum_m Z^{\frac{\lambda}{m}}(z)\left|(\lambda,\mu)I = \frac{\lambda}{2}(M_I - m)\right\rangle\left\langle\frac{\lambda}{2}(M_I - m)\ \frac{w}{2}m\middle|IM_I\right\rangle.$$

From now on the square bracket will denote angular momentum coupling in the order indicated; ie a right to left coupling order. (This order, although somewhat unconventional, will prove very convenient in avoiding angular momentum reordering phase factors).

Since the above is an orthonormal basis with respect to the Bargmann z-space measure, the nonunitary form of the generators, $\Gamma(E_{i3})$ and $\Gamma(E_{3i})$, must still be converted to unitary form. This can again be achieved by a similarity transformation with a K-operator. Since the "collective" angular momentum $\frac{w}{2}$ is uniquely determined by the $U(1)$ quantum number $Y = \frac{1}{3}(\lambda + 2\mu) - w$, the K-operation in this more challenging example is still only the simple command: Multiply by a normalization factor. The operator K can therefore again be made hermitian, $K^\dagger = K$. Since the $\Gamma(O)$, for O an operator of the $U(1) + SU(2)$ subalgebra, are already unitary with respect to the Bargmann measure, the K operator can be chosen to be an isoscalar, Y-invariant operator; and its matrix elements will be diagonal in M_I and independent of M_I. The unitary form of the z-space operators will again be designated by γ,

$$\gamma(E_{3i}) = K^{-1}\Gamma(E_{3i})K; \qquad \gamma(E_{i3}) = K^{-1}\Gamma(E_{i3})K;$$

where

$$\gamma(E_{3i}) = (\gamma(E_{i3}))^\dagger = (K^{-1}\partial_i K)^\dagger = Kz_iK^{-1} = K^{-1}\Gamma(E_{3i})K.$$

Right and left multiplication by K again converts this to an equation for the determination of K^2

$$\Gamma(E_{3i})K^2 = K^2 z_i.$$

This will again be solved by the introduction of the auxiliary operator, $\Lambda_{op.}$, with the property

$$[\Lambda_{op.}, z_i] = \Gamma(E_{3i}) = z_\alpha E_{\alpha i} - z_i E_{33} - z_i z_\alpha \partial_\alpha.$$

This has the solution

$$\Lambda_{op.} = z_\alpha E_{\alpha\beta}\partial_\beta - E_{33}(z_\alpha\partial_\alpha) - \frac{1}{2}(z_\alpha\partial_\alpha)(z_\beta\partial_\beta) + \frac{1}{2}(z_\alpha\partial_\alpha)$$

$$= -2I^{intr.}\cdot I^{coll.} + (\frac{1}{2} + \frac{3}{2}Y)(z_\alpha\partial_\alpha) - \frac{1}{2}(z_\alpha\partial_\alpha)(z_\beta\partial_\beta).$$

Since $z_\alpha\partial_\alpha$ has the simple eigenvalue w, and the eigenvalue of the scalar product $I^{intr.}\cdot I^{coll.}$ is given in terms of the quantum numbers, I, $\frac{1}{2}\lambda$, $\frac{1}{2}w$ in the usual way, the eigenvalue of Λ follows

$$\Lambda_{w,I} = -I(I+1) + \frac{\lambda}{2}(\frac{\lambda}{2} + 1) + \frac{w}{2}(\frac{w}{2} + 1) + (\frac{1}{2} + \frac{1}{2}(\lambda + 2\mu))w - \frac{1}{2}w^2.$$

The equation for the determination of K^2 can now be put into the form

$$(\Lambda_{op.}z_i - z_i\Lambda_{op.})K^2 = K^2 z_i.$$

Taking matrix elements of this relation between states with quantum numbers w, I on the right and $w+1$, $I' = (I \pm \frac{1}{2})$ on the left, leads to the recursive equation

$$(\Lambda_{w+1,I'} - \Lambda_{w,I})K^2_{w,I} = K^2_{w+1,I'},$$

where the matrix element of z_i has been cancelled since it is common to both sides of the equation. The K^2-ratio then follows from

$$\Lambda_{w+1,I+\frac{1}{2}} - \Lambda_{w,I} = \frac{1}{2}(-w + \lambda + 2\mu) - I$$

$$\Lambda_{w+1,I-\frac{1}{2}} - \Lambda_{w,I} = \frac{1}{2}(-w + \lambda + 2\mu) + I + 1.$$

Since action with the creation type operators increases the degree of the z-space polynomial by 1 unit and changes the value of the isospin by $\pm\frac{1}{2}$, repeated action with w such operators can be made up of p isospin lowering and q isospin raising steps, with

$$w = p + q; \qquad I = \frac{1}{2}\lambda - \frac{1}{2}p + \frac{1}{2}q.$$

In terms of the labels p and q the K^2-ratios have the simpler values

$$\frac{K^2_{w+1,I+\frac{1}{2}}}{K^2_{w,I}} = \frac{K^2_{p,q+1}}{K^2_{p,q}} = \Lambda_{w+1,I+\frac{1}{2}} - \Lambda_{w,I} = (\mu - q)$$

$$\frac{K^2_{w+1,I-\frac{1}{2}}}{K^2_{w,I}} = \frac{K^2_{p+1,q}}{K^2_{p,q}} = \Lambda_{w+1,I-\frac{1}{2}} - \Lambda_{w,I} = (\lambda + \mu + 1 - p).$$

Repeated application, starting with $K^2_{p=0,q=0} = 1$ for the normalized starting states, leads to

$$K^2_{p,q} = \frac{\mu!(\lambda + \mu + 1)!}{(\mu - q)!(\lambda + \mu + 1 - p)!}.$$

With the knowledge of K the matrix elements of the unitary form of the creation type operators $E_{3i} = A_i$ between orthonormal Bargmann basis states, $\chi(z)$, can now be evaluated

$$\langle \chi_{[\frac{\lambda}{2}\times\frac{w+1}{2}]I'}\|\gamma(\mathbf{A})\|\chi_{[\frac{\lambda}{2}\times\frac{w}{2}]I}\rangle = \langle \chi_{[\frac{\lambda}{2}\times\frac{w+1}{2}]I'}\|KzK^{-1}\|\chi_{[\frac{\lambda}{2}\times\frac{w}{2}]I}\rangle.$$

Since these are representation independent, they also lead to

$$\langle [\frac{\lambda}{2}\times\frac{w+1}{2}]I'\|\mathbf{A}\|[\frac{\lambda}{2}\times\frac{w}{2}]I\rangle = \frac{K_{w+1,I'}}{K_{w,I}}\langle \chi_{[\frac{\lambda}{2}\times\frac{w+1}{2}]I'}\|z\|\chi_{[\frac{\lambda}{2}\times\frac{w}{2}]I}\rangle$$

$$= \frac{K_{w+1,I'}}{K_{w,I}}\,U\left(\frac{\lambda}{2}\frac{w}{2}I'\frac{1}{2};I\frac{w+1}{2}\right)\left\langle\frac{w+1}{2}\|z\|\frac{w}{2}\right\rangle,$$

where we have used standard angular momentum recoupling theory to evaluate the reduced matrix element of z, a rank $\frac{1}{2}$ operator acting only in the collective space, in a basis in which the intrinsic angular momentum $\frac{\lambda}{2}$ is coupled with the collective angular momentum $\frac{w}{2}$ to resultant I:

$$\langle \chi_{[\frac{\lambda}{2}\times\frac{w+1}{2}]I'}\|z\|\chi_{[\frac{\lambda}{2}\times\frac{w}{2}]I}\rangle = \left\langle \chi_{[\frac{\lambda}{2}\times\frac{w+1}{2}]I'M'_I}\Big|\left[z^{\frac{1}{2}}\times\chi_{[\frac{\lambda}{2}\times\frac{w}{2}]I}\right]^{I'}_{M'_I}\right\rangle$$

$$= \left\langle \chi_{[\frac{\lambda}{2}\times\frac{w+1}{2}]I'M'_I}\Big|\left[z^{\frac{1}{2}}\times\left[Z^{\frac{w}{2}}(z)\times|\frac{\lambda}{2}\rangle\right]^I\right]^{I'}_{M'_I}\right\rangle$$

$$= U\left(\frac{\lambda}{2}\frac{w}{2}I'\frac{1}{2};I\frac{w+1}{2}\right)\left\langle\frac{w+1}{2}\|z\|\frac{w}{2}\right\rangle.$$

The U-coefficient, $U(j_1 j_2 j j_3; j_{12} j_{23}) = \sqrt{(2j_{12}+1)(2j_{23}+1)}W(j_1 j_2 j j_3; j_{12} j_{23})$ is a Racah W-coefficient in unitary form. It could also have been obtained from the ratio of the reduced matrix element in the fully coupled [intr. × coll.] → full basis relative to the reduced matrix element in the purely *collective* basis. This ratio is given by a $9-j$ recoupling coefficient, (again in unitary form),

$$\begin{bmatrix} \frac{\lambda}{2} & \frac{w}{2} & I \\ 0 & \frac{1}{2} & \frac{1}{2} \\ \frac{\lambda}{2} & \frac{w+1}{2} & I' \end{bmatrix}$$

where this $9 - j$ coefficient, with entry 0 in the number 3 position, is the above U-coefficient. This U-coefficient, with one angular momentum of $\frac{1}{2}$, is extremely simple. It has the values

$$\sqrt{\frac{(\frac{\lambda}{2} + \frac{w}{2} + I + 2)(-\frac{\lambda}{2} + \frac{w}{2} + I + 1)}{(2I' + 1)(w + 1)}} \qquad \text{for} \quad I' = I + \frac{1}{2}$$

$$\sqrt{\frac{(\frac{\lambda}{2} + \frac{w}{2} - I + 1)(\frac{\lambda}{2} - \frac{w}{2} + I)}{(2I' + 1)(w + 1)}} \qquad \text{for} \quad I' = I - \frac{1}{2}$$

The angular momentum reduced matrix elements of the creation-type operators, $A_i = E_{3i}$, are thus

$$\left\langle [\frac{\lambda}{2} \times \frac{w + 1}{2}] I' \| \mathbf{A} \| [\frac{\lambda}{2} \times \frac{w}{2}] I \right\rangle$$

$$= \sqrt{(\mu - q)} \sqrt{\frac{(\lambda + q + 2)(q + 1)}{(2I' + 1)}} \qquad \text{for} \qquad I' = I + \frac{1}{2}; \quad p, q \to p, q + 1$$

$$= \sqrt{(\lambda + \mu + 1 - p)} \sqrt{\frac{(\lambda - p)(p + 1)}{(2I' + 1)}} \qquad \text{for} \qquad I' = I - \frac{1}{2}; \quad p, q \to p + 1, q$$

where the first factor comes from the K-value ratios, and the second factor is the U-coefficient $\times \sqrt{w + 1}$. The reduced matrix elements of the $SU(3)$ generators are thus determined essentially by a recoupling coefficient of the simpler subgroup $SU(2)$.

The range of the p, q values, which determines the possible Y, I combinations is given by

$$0 \le q \le \mu ; \qquad\qquad 0 \le p < \lambda$$

where the first follows from the K-ratios, and the second from the value of the Racah coefficients. With these limits, the possible Y, I-values can be displayed on a diamond-shaped diagram. With $\mu \le \lambda$, eg, the starting values, $Y = \frac{1}{3}(\lambda + 2\mu)$, $I = \frac{1}{2}\lambda$ form the lowest point of the diamond. After $w = \mu$ $(1, 2 \to 3$ conversion$)$ or "raising" steps, with Y having reached the value $\frac{1}{3}(\lambda - \mu)$, the I-spin can range from $\frac{1}{2}(\lambda - \mu)$ to $\frac{1}{2}(\lambda + \mu)$, where the last value makes the right corner of the diamond. After a total of $w = \lambda$ "raising" steps, Y has the value $\frac{2}{3}(-\lambda + \mu)$, and I ranges from 0, the left corner of the diamond, to μ. Finally, after a total of $w = (\lambda + \mu) = w_{max}$. "raising" steps, the top of the diamond is reached, with $Y = -\frac{1}{3}(2\lambda + \mu)$ and the unique value $I = \frac{1}{2}\mu$; see Fig. 3.

With the determination of the $K_{p,q}$ or $K_{w,I}$ normalization factors, it is also possible to express the normalized states in terms of the action of w raising operators, $\mathbf{A}$, acting in succession on the starting states. By inverting

$$\gamma(\mathbf{A}) \Longrightarrow K z K^{-1}$$

the polynomial $Z^{\frac{1}{2}w}(\mathbf{z})$ can be converted to a polynomial in $\mathbf{A}$ via

$$\mathbf{z} \times \mathbf{z} \cdots \times \mathbf{z} \times \mathbf{z} \Longrightarrow K^{-1}\mathbf{A}KK^{-1}\mathbf{A}K \cdots K^{-1}\mathbf{A}KK^{-1}\mathbf{A}K.$$

With the cancellation of interior $K\,K^{-1}$ factors, and the value $K_{p=0,q=0} = 1$ for the extreme right hand factor, acting on the normalized starting states, this converts

$$\left[Z^{\frac{1}{2}w}(\mathbf{z}) \times \left|\frac{\lambda}{2}\right\rangle\right]^I_{M_I} \Longrightarrow K^{-1}_{w,I}\left[Z^{\frac{1}{2}w}(\mathbf{A}) \times \left|\frac{\lambda}{2}\right\rangle\right]^I_{M_I},$$

where the $\mathbf{z}'s$ are replaced by $\mathbf{A}'s$ in the functions Z. An orthonormal basis can therefore be constructed in terms of raising operators $\mathbf{A}$ by

$$\left|[\frac{\lambda}{2} \times \frac{w}{2}]IM_I\right\rangle = K^{-1}_{w,I}\left[Z^{\frac{1}{2}w}(\mathbf{A}) \times \left|\frac{\lambda}{2}\right\rangle\right]^I_{M_I}.$$

This construction also proves very useful in the evaluation of $SU(3)$ Wigner coefficients. (The evaluation of the Wigner coefficients involving the coupling of an arbitrary representation with the fundamental representation, $(\lambda, \mu) = (10)$ will be illustrated in Chapter 5).

Finally, the reduced matrix elements of the annihilation type operators, $B_i = E_{i3}$, follow from hermitian conjugation in the usual way

$$\left\langle [\frac{\lambda}{2} \times \frac{w}{2}]I \|B\| [\frac{\lambda}{2} \times \frac{w+1}{2}]I' \right\rangle$$

$$= \sqrt{\frac{(2I'+1)}{(2I+1)}}(-1)^{I+\frac{1}{2}-I'}\left\langle [\frac{\lambda}{2} \times \frac{w+1}{2}]I' \|A\| [\frac{\lambda}{2} \times \frac{w}{2}]I \right\rangle.$$

The full matrix representations of the $SU(3)$ algebra have thus been achieved.

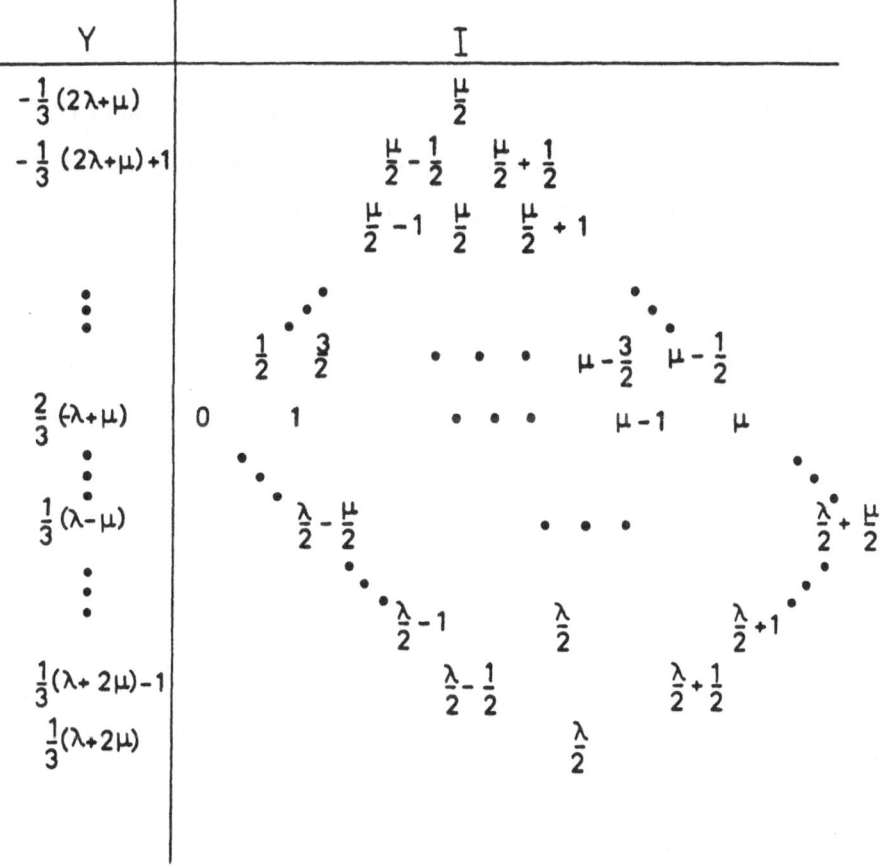

Figure 3

3.2 The Neutron-proton Quasispin Algebra: A Simple Fermion Pair Algebra.

As our second detailed example we shall consider another algebra with a simple $SU(2) + U(1)$ subalgebra, the so-called neutron-proton quasispin algebra. Despite similarities with the SU(3) algebra, this example will illustrate a new feature, shared with most other applications of the vector coherent state method. It is our first example where the K operator involves more than the simple command: multiply with a normalization factor.

Since the neutron-proton quasispin algebra is a special case of a fermion pair algebra, a few general words about such algebras may be useful.

General Fermion Pair Algebras:

If $b_i^\dagger$, b_i are fermion creation and annihilation operators, where the label, i now stands for all the quantum numbers, space, spin, isospin, (color or others, if needed), the algebra bilinear in the $b_i^\dagger$, b_i falls into our general category. With

$$
\begin{array}{lll}
A_{ik} = -A_{ki} = b_i^\dagger b_k^\dagger & i, k = 1, ..., N & \text{"Creation" Ops.} \\
B_{ik} = -B_{ki} = b_k b_i & i, k = 1, ..., N & \text{"Annihilation" Ops.} \\
C_{ik} = (b_i^\dagger b_k - \frac{1}{2}\delta_{ik}) & i, k = 1, ..., N & U(N) \text{ Generators}
\end{array}
$$

these operators satisfy the commutation relations

$$[B_{ij}, B_{k\ell}] = 0$$

$$[B_{ij}, A_{k\ell}] = \delta_{jk}C_{\ell i} + \delta_{i\ell}C_{kj} - \delta_{ik}C_{\ell j} - \delta_{j\ell}C_{ki}$$

$$[B_{ij}, C_{k\ell}] = \delta_{ik}B_{j\ell} + \delta_{jk}B_{\ell i}$$

and satisfy our criteria 1) to 4). The vacuum is now the true fermion vacuum, (or possibly a single-fermion state, $|k\rangle = b_k^\dagger|0\rangle$). The core subalgebra, $U(N)$, contains the Cartan subalgebra of the full algebra. From the commutation relations we can associate the A_{ik}, B_{ik}, C_{ik} with $2N$-dimensional angular momentum operators

$$A_{ik} = \frac{1}{2}(J_{2i-1,2k} + J_{2i,2k-1} + iJ_{2i,2k} - iJ_{2i-1,2k-1})$$

$$B_{ik} = \frac{1}{2}(J_{2i-1,2k} + J_{2i,2k-1} - iJ_{2i,2k} + iJ_{2i-1,2k-1})$$

$$C_{ik} = \frac{1}{2}(J_{2i-1,2k} - J_{2i,2k-1} + iJ_{2i,2k} + iJ_{2i-1,2k-1})$$

so that we can see that the above $(2N-1)N$ operators generate an $SO(2N)$ algebra.

For a realistic fermion system, eg a system of nucleons in a nucleus with several active subshells, j, the detailed application of the vector coherent state method may reduce the problem to fairly standard spectroscopy. However, this will in general require detailed knowledge of a complicated $U(N)$ subalgebra, with large N. On the other hand, if the fermion quantum numbers, i, can be split naturally into two sets, $i \equiv \alpha k$, fermion pair operators, such as $\sum_\alpha b^\dagger_{\alpha k} b^\dagger_{-\alpha j}$, and their partners, may generate a simpler algebra with interesting applications.

The neutron-proton quasispin algebra is such an algebra. For a single j-shell it is generated by the $J = 0, T = 1$-pair creation operators

$$A(M_T) = \frac{1}{2}\sum_m \sum_{m_{t_1}, m_{t_2}} (-1)^{j-m} \langle \frac{1}{2}m_{t_1} \; \frac{1}{2}m_{t_2}|1 \; M_T\rangle b^\dagger_{jmm_{t_1}} b^\dagger_{j-mm_{t_2}}$$

For mixed configurations a sum over active j-shells can be included. The family of 10 operators built from the 3 $J = 0, T = 1$-pair creation operators, $A(M_T)$, the 3 hermitian conjugate pair annihilation operators, $B(M_T)$, and the isospin, nucleon number subgroup, $SU(2) + U(1)$, satisfy our criteria 1) to 4). Explicitly, they are:

$$A(1) = \sum_{m>0}(-1)^{j-m}a^\dagger_{jm\frac{1}{2}}a^\dagger_{j-m\frac{1}{2}} \implies E_{14}$$

$$A(0) = \frac{1}{\sqrt{2}}\sum_{m>0}(-1)^{j-m}(a^\dagger_{jm\frac{1}{2}}a^\dagger_{j-m-\frac{1}{2}} + a^\dagger_{jm-\frac{1}{2}}a^\dagger_{j-m\frac{1}{2}}) \implies \frac{1}{\sqrt{2}}(E_{13} - E_{24})$$

$$A(-1) = \sum_{m>0}(-1)^{j-m}a^\dagger_{jm-\frac{1}{2}}a^\dagger_{j-m-\frac{1}{2}} \implies E_{23}$$

$$B(1) = \sum_{m>0}(-1)^{j-m}a_{j-m\frac{1}{2}}a_{jm\frac{1}{2}} \implies E_{41}$$

$$B(0) = \frac{1}{\sqrt{2}}\sum_{m>0}(-1)^{j-m}(a_{j-m-\frac{1}{2}}a_{jm\frac{1}{2}} + a_{j-m\frac{1}{2}}a_{jm-\frac{1}{2}}) \implies \frac{1}{\sqrt{2}}(E_{31} - E_{42})$$

$$B(-1) = \sum_{m>0}(-1)^{j-m}a_{j-m-\frac{1}{2}}a_{jm-\frac{1}{2}} \implies E_{32}$$

$$T_+ = \sum_m a^\dagger_{jm\frac{1}{2}}a_{jm-\frac{1}{2}} \implies (E_{12} + E_{34})$$

$$T_0 = \frac{1}{2}\sum_m(a^\dagger_{jm\frac{1}{2}}a_{jm\frac{1}{2}} - a^\dagger_{jm-\frac{1}{2}}a_{jm-\frac{1}{2}}) \implies \frac{1}{2}(E_{11} - E_{22} + E_{33} - E_{44})$$

$$T_- = \sum_m a^\dagger_{jm-\frac{1}{2}}a_{jm\frac{1}{2}} \implies (E_{21} + E_{43})$$

$$\frac{1}{2}N_{op.} - \Omega = \frac{1}{2}\sum_{m,m_t} a^\dagger_{jmm_t}a_{jmm_t} - \Omega \implies \frac{1}{2}(E_{11} + E_{22} - E_{33} - E_{44})$$

where $\Omega = (j + \frac{1}{2})$ for a single j-shell. Alternately, $\Omega = \sum_j(j + \frac{1}{2})$ for a mixed configuration of several j-shells. It is useful to transform to "cartesian" coordinates:

$$A(\pm 1) = \mp\frac{1}{\sqrt{2}}(A_x \pm iA_y), \qquad A(0) = A_z;$$

$$B(\pm 1) = \mp\frac{1}{\sqrt{2}}(B_x \pm iB_y), \qquad B(0) = B_z;$$

$$and \qquad T_\pm = (T_x \pm iT_y), \qquad T_0 = T_z.$$

In terms of these cartesian components, the operators satisfy the commutation relations

$$[B_k, A_j] = i\epsilon_{kjl}T_l - \delta_{kj}(\frac{1}{2}N_{op.} - \Omega);$$

$$[B_k, T_j] = i\epsilon_{kjl}B_l; \qquad [B_k, \frac{1}{2}N_{op.} - \Omega] = B_k.$$

By making the identification

$$A_k = \frac{1}{\sqrt{2}}(J_{k4} + iJ_{k5}); \qquad B_k = \frac{1}{\sqrt{2}}(J_{k4} - iJ_{k5});$$

$$\mathbf{T} = (J_{23}, J_{31}, J_{12}); \qquad \frac{1}{2}N_{op.} - \Omega = J_{45};$$

we see that the neutron-proton quasispin algebra can be put into correspondence with an $SO(5)$ algebra, where our core subalgebra is an $SO(3)+SO(2)$ subalgebra generated by J_{ik}, with $i,k = 1,2,3$, and J_{45}. The well known isomorphism between $SO(5)$ and the unitary symplectic algebra, $Sp(4)$, is indicated above where the correspondence between the $A(M_T), B(M_T), \mathbf{T}$, and $U(4)$ generators E_{ik} has been shown (by means of $\Longrightarrow$). The fact that the linear combinations of the E_{ik} which occur generate the subalgebra $Sp(4)$ of $U(4)$ can be seen most easily if we make the correspondence

$$(1,2,3,4) \quad \Longrightarrow \quad (+\frac{3}{2}, +\frac{1}{2}, -\frac{1}{2}, -\frac{3}{2}) \equiv (m_i)$$

The combinations of the E_{ik} which belong to the above family of 10 operators are of the form

$$\frac{1}{\sqrt{2(1 + \delta_{m_1 m_2})}}(E_{m_1, -m_2} + (-1)^{m_1 - m_2} E_{m_2, -m_1}).$$

These are therefore $Sp(4)$ generators.

The vacuum states are the states annihilated by the $J = 0$, $T = 1$-pair annihilation operators, $B(M_T)$. They are the states entirely free of $J = 0, T = 1$ pairs, with nucleon number, $n = v$, and isospin, t, where $v =$ seniority number, and t = reduced isospin, the isospin of the v nucleons entirely free of $J = 0$, $T = 1$ pairs:

$$B(M_T)|v, tm_t\rangle = 0 \qquad for\ all\ M_T,\ m_t.$$

The J-content of nuclear shell model states, carrying the details of the space-spin characteristics of many-nucleon states, is determined solely by the quantum numbers v and t of these "vacuum" states, (and of course the j-value or the mixture of j-values of the active shells). The final n and T combinations of the many-nucleon states are determined by the successive actions of the creation operators, $A(M_T)$. The neutron-proton quasispin formalism thus gives us a simple means for extracting the (n,T)-dependent factors of shell model matrix elements in the seniority scheme. Although earlier (laborious!) algebraic methods have achieved this for the simpler states of low seniority, $v \leq 2$, the vector coherent state technique permits an easy extension to states of arbitrarily high seniority, so that the shell model problem for n nucleons has truly been reduced to one for v nucleons, even if admixtures of high-v states become important.

Since the $J = 0$, $T = 1$-pair creation operators have three components, the vector coherent state is built in terms of three complex variables, $(z_1, z_2, z_3) = \mathbf{z}$. It is expressed in the simplest way in terms of the cartesian components, A_i, of the pair creation operators. With $\mathbf{z^*} \cdot \mathbf{A} = z_1^* A_1 + z_2^* A_2 + z_3^* A_3$:

$$|\mathbf{z}\rangle = e^{\mathbf{z^*} \cdot \mathbf{A}}|v, tm_t\rangle$$

State vectors are again mapped into z-space functions

$$|\psi\rangle \Longrightarrow \psi_{v, tm_t}(\mathbf{z}) = \langle v, tm_t|e^{\mathbf{z} \cdot \mathbf{B}}|\psi\rangle,$$

and operators into their z-space realizations

$$O|\psi\rangle \Longrightarrow \Gamma(O)\psi_{v, tm_t}(\mathbf{z}) = \langle v, tm_t|\left(O + [\mathbf{z} \cdot \mathbf{B}, O] + \frac{1}{2}[\mathbf{z} \cdot \mathbf{B}, [\mathbf{z} \cdot \mathbf{B}, O]] + \cdots\right)e^{\mathbf{z} \cdot \mathbf{B}}|\psi\rangle$$

The annihilation operators, B_i, (in cartesian form), are mapped into simple z-space derivative operators, where we will use the shorthand notation, $\partial/\partial z_i = \nabla_i$.

The operators $\frac{1}{2}N_{op.} - \Omega$ and T_i are mapped into an intrinsic part, which arises from the operators O acting to the left on the vacuum states, and a purely z-dependent part. The latter arises from the first commutator terms, such as $[\mathbf{z}\cdot\mathbf{B}, T_i] = iz_j\epsilon_{jik}B_k \longrightarrow -i\epsilon_{ijk}z_j\nabla_k$, where such terms can then be expressed in terms of standard vector notation. The operator $N_{op.}$ acting on the vacuum states counts the number of nucleons free of $J = 0$, $T = 1$-coupled pairs, $N^{intr.} = v$. The components of $\mathbf{T}$, acting on the v-nucleon states of reduced isospin, t, will again be denoted by $\mathbf{T}^{intr.} = \not{t}$. The pair creation operators, A_i, which, when acting to the left, annihilate the vacuum states, again get contributions only from the single and double commutator terms

$$[\mathbf{z}\cdot\mathbf{B}, A_i] = -i\epsilon_{ijk}z_jT_k - z_i(\frac{1}{2}N_{op.} - \Omega)$$

$$[\mathbf{z}\cdot\mathbf{B}, [\mathbf{z}\cdot\mathbf{B}, A_i]] = z_jz_jB_i - 2z_iz_jB_j$$

When expressed in terms of standard vector notation, the z-space realizations of the 10 generators of the neutron-proton quasispin algebra are then

$$\Gamma(\mathbf{B}) = \nabla$$

$$\Gamma(\mathbf{T}) = \not{t} - i[\mathbf{z} \times \nabla] \qquad = \mathbf{T}^{intr.} + \mathbf{T}^{coll.}$$

$$\Gamma(\frac{1}{2}N_{op.} - \Omega) = \frac{1}{2}v - \Omega + \mathbf{z}\cdot\nabla \qquad = \frac{1}{2}N^{intr.} - \Omega + \frac{1}{2}N^{coll.}$$

$$\Gamma(\mathbf{A}) = -i[\mathbf{z} \times \not{t}] - \mathbf{z}(\frac{1}{2}v - \Omega) + \frac{1}{2}(\mathbf{z}\cdot\mathbf{z})\nabla - \mathbf{z}(\mathbf{z}\cdot\nabla).$$

We have therefore mapped the complicated algebra into a direct sum of a simple oscillator algebra and an intrinsic $SU(2)\mathrm{x}U(1)$ algebra. The main difference from our previous example is that the oscillator algebra is now a 3-dimensional one, generated by z_i, and $\partial/\partial z_i$, with $i = 1, 2, 3$. The components of $\mathbf{z}$ and ∇ again commute with the intrinsic operators

$$[\mathbf{z}, \not{t}] = [\nabla, \not{t}] = [\mathbf{z}, N^{intr.}] = [\nabla, N^{intr.}] = 0.$$

For the sake of uniformity of notation, we have again named the purely z-dependent parts of the operators $\mathbf{T}$ and $N_{op.}$ as the "collective" parts of these operators.

In constructing an orthonormal z-space oscillator basis, it will now be convenient to make the transformation from the cartesian normalized functions, $\frac{z_1^a z_2^b z_3^c}{\sqrt{a!b!c!}}$, of degree $p = a + b + c$, to a basis of z-space functions of good angular momentum T_p, M_{T_p}. For the 3-dimensional harmonic oscillator this transformation is well known, from the single particle functions of the nuclear shell model, eg. The functions of degree, p, belong to the totally symmetric (single-rowed Young tableau) representations of SU(3), with $(\lambda, \mu) = (p0)$, using the Elliott notation of the last section. For these totally symmetric representations, the possible values of T_p are $p, p - 2, ..., 0$(or 1) for $p =$even(or odd), again a result well known in nuclear physics from the properties of single particle shell model functions. (Note that we now need eigenfunctions reduced not according to the $SU(3) \supset SU(2) \times U(1)$ chain, discussed in terms of an intrinsic x', y', z' coordinate system in the last section, but reduced according to the $SU(3) \supset SO(3)$ chain of good angular momentum). Although these angular momentum eigenfunctions are complicated for general $SU(3)$ representations, (λ, μ), they are very simple for the totally symmetric representations, $(p0)$, of a single 3-dimensional oscillator variable $\mathbf{z}$. When expressed in terms of oscillator creation operators, the normalization factors of such functions are also well known (Moshinsky, 1961), so that we can write down the orthonormal z-space functions of degree p and good "collective" isospin, T_p:

$$Z_{T_p, M_{T_p}}^{(p0)}(\mathbf{z}) = \sqrt{\frac{4\pi 2^{T_p}[\frac{1}{2}(p + T_p)]!}{[\frac{1}{2}(p - T_p)]!(p + T_p + 1)!}} (\mathbf{z} \cdot \mathbf{z})^{\frac{1}{2}(p - T_p)} Y_{T_p, M_{T_p}}(\mathbf{z}),$$

where the $Y_{T_p,M_{T_p}}$ are normalized solid harmonics; *eg*

$$Y_{1m}(z) = \sqrt{\frac{3}{4\pi}}(z_{+1}, z_0, z_{-1}) \qquad with \quad z_{\pm 1} = \mp \frac{1}{\sqrt{2}}(z_1 \pm iz_2), \ z_0 = z_3.$$

For most of our purposes it is sufficient to know the explicit form of the states with $M_{T_p} = T_p$:

$$Z_{T_p,T_p}^{(p0)}(z) = N_{p,T_p}(z \cdot z)^{\frac{1}{2}(p-T_p)}\left(-\frac{1}{\sqrt{2}}(z_1 + iz_2)\right)^{T_p} \qquad with$$

$$N_{p,T_p} = \sqrt{\frac{(2T_p + 1)![\frac{1}{2}((p + T_p)]!}{[\frac{1}{2}(p - T_p)]!(p + T_p + 1)![(T_p)!]^2}}.$$

From these the angular momentum-reduced matrix element of z in its own "collective" z-space, $\langle (p+1,0)T_p'\|z\|(p0)T_p\rangle$, can be evaluated at once. For $T_p' = (T_p+1)$ this reduced matrix element follows from the ratio of normalization constants, $N_{p+1,T_p+1}/N_{p,T_p}$, after multiplication of $Z_{T_p,T_p}^{(p0)}$ with z_{+1}. For $T_p' = (T_p - 1)$ it follows from

$$\langle (p+1,0)T_p - 1\|z\|(p0)T_p\rangle = \sqrt{\frac{(2T_p + 1)}{(2T_p - 1)}}\langle (p0)T_p\|\frac{\partial}{\partial z}\|(p+1,0)T_p - 1\rangle,$$

and

$$\frac{1}{\sqrt{2}}\left(\frac{\partial}{\partial z_1} + i\frac{\partial}{\partial z_2}\right)N_{p+1,T_p-1}(z \cdot z)^{\frac{1}{2}(p-T_p+2)}\left(-\frac{1}{\sqrt{2}}(z_1 + iz_2)\right)^{T_p-1}$$

$$= -N_{p+1,T_p-1}(p - T_p + 2)(z \cdot z)^{\frac{1}{2}(p-T_p)}\left(-\frac{1}{\sqrt{2}}(z_1 + iz_2)\right)^{T_p}.$$

These relations lead to the reduced matrix elements

$$\langle (p+1,0)T_p + 1\|z\|(p0)T_p\rangle = \sqrt{\frac{(p + T_p + 3)(T_p + 1)}{(2T_p + 3)}}$$

$$\langle (p+1,0)T_p - 1\|z\|(p0)T_p\rangle = -\sqrt{\frac{T_p(p - T_p + 2)}{(2T_p - 1)}}$$

The vector, z, has the same transformation properties as the vector, $\mathbf{A}$. It transforms according to the 3-dimensional $SU(3)$ representation, $(\lambda, \mu) = (10)$.

The above reduced matrix elements could therefore also have been obtained from known $SU(3) \supset SO(3)$ reduced (double-barred) Wigner coefficients through

$$\langle (p+1,0)T'_p \| z \| (p0)T_p \rangle = \langle (p0)\ell = T_p; (10)\ell_0 = 1 \| (p+1,0)\ell' = T'_p \rangle \sqrt{(p+1)},$$

where the oscillator factor, $\sqrt{(p+1)}$, is the $SU(3)$-reduced matrix element of z which follows from the composition of $SU(3)$- coupled z-space functions

$$\left[Z^{(p0)}(z) \times z^{(10)} \right]^{(p+1,0)} = \sqrt{(p+1)} Z^{(p+1,0)}(z), \qquad \text{or}$$

$$\left[Z^{(p_1 0)}(z) \times Z^{(p_2 0)}(z) \right]^{(p0)} = \delta_{p,p_1+p_2} \sqrt{\frac{p!}{p_1! p_2!}} Z^{(p0)}(z).$$

(Note that the square brackets here denote $SU(3)$-coupling). In terms of these "collective" z-space functions, it is straightforward to construct a full orthonormal z-space basis in terms of the angular-momentum coupled functions

$$\left[Z^{(p0)}_{T_p}(z) \times |v,t\rangle \right]_{TM_T}$$

where the square bracket now denotes angular momentum coupling, again in the right to left coupling-order convention, corresponding to $[t \times T_p] \to T$. Since this z-space basis is orthonormal with respect to a scalar product with the Bargmann measure, and since the z-space realizations, $\Gamma(\mathbf{A})$ and $\Gamma(\mathbf{B})$, are not unitary with respect to this measure, it is again necessary to transform these to a unitary realization. We will again attempt to do this with a hermitian operator, $K = K^\dagger$, through

$$\gamma(\mathbf{A}) = K^{-1}\Gamma(\mathbf{A})K; \quad \gamma(\mathbf{B}) = K^{-1}\Gamma(\mathbf{B})K$$

Since the $\Gamma(\mathbf{T})$ are already unitary, we can set

$$\gamma(\mathbf{T}) = \Gamma(\mathbf{T});$$

and similarly for $N_{op.}$. Thus K will commute with $\mathbf{T}$ and $N_{op.}$ and will be a number-preserving T-invariant quantity independent of M_T. As in the earlier examples, the requirement of unitarity

$$\gamma(A_i) = (\gamma(B_i))^\dagger = (K^{-1}\frac{\partial}{\partial z_i}K)^\dagger = Kz_iK^{-1} = K^{-1}\Gamma(A_i)K$$

leads by left and right multiplication with K to the relation

$$\Gamma(A_i)K^2 = K^2 z_i$$

from which K^2 can be determined.

By inverting the relation

$$\gamma(\mathbf{A}) = K\mathbf{z}K^{-1}$$

we can again convert the symmetric z-space polynomial, $Z^{(p0)}(\mathbf{z})$, into the corresponding $Z^{(p0)}(\mathbf{A})$ via

$$\mathbf{z} \times \mathbf{z} \times \cdots \times \mathbf{z} = K^{-1}\mathbf{A}KK^{-1}\mathbf{A}KK^{-1}\cdots KK^{-1}\mathbf{A}K$$

to lead to the transformation

$$\left[Z^{(p0)}_{T_p}(\mathbf{z}) \times |v,t\rangle\right]_{TM_T} \Longrightarrow K^{-1}\left[Z^{(p0)}_{T_p}(\mathbf{A}) \times |v,t\rangle\right]_{TM_T}$$

where the operator K^{-1} which converts

$$\left[Z^{(p0)}_{T_p}(\mathbf{A}) \times |v,t\rangle\right]_{TM_T} \equiv |\Psi(vt; pTM_T; T_p)\rangle$$

into an orthonormal basis is now more complicated than in the previous examples. The matrix elements of K are diagonal in v,t, since these are bona fide quantum numbers associated with the $Sp(4)$ irreducible representations, and are also diagonal in $p = \frac{1}{2}(n-v), T, M_T$, since these are quantum numbers associated with the $U(1) \times SU(2)$ subgroup. Note, however, that the label, T_p, differs from these bona fide quantum numbers. Despite the physical significance of this label, there is now no hermitian operator with eigenvalues naturally associated with T_p. In general, there is more than one possible value of T_p for fixed t, T, p. Eg, with $t = 1, T = 1$, and $p = 2$, *both* $T_p = 0$ *and* $T_p = 2$ are possible values of this label. (Note the difference from the previous example of the $SU(3) \supset SU(2) \times U(1)$ group where the "collective" isospin, $\frac{w}{2}$, was uniquely fixed by the quantum number, w, which is itself determined by the hypercharge, Y, *ie* the $U(1)$ quantum number).

As a result, the states $|\Psi\rangle$, generated by the action of $Z_{T_p}^{(p0)}(\mathbf{A})$ on the vacuum states, unlike the orthonormal z-space states

$$\left[Z_{T_p}^{(p0)}(\mathbf{z}) \times |v, t\rangle \right]_{TM_T} \equiv \chi_{vt;pTM_T;T_p}(\mathbf{z}),$$

now lead to the scalar product

$$\left\langle \Psi(vt; pTM_T; T_p') \middle| \Psi(vt; pTM_T; T_p) \right\rangle$$

$$= \frac{1}{\pi^3} \int d^2 z_1 \int d^2 z_2 \int d^2 z_3 \ e^{-(\mathbf{z} \cdot \mathbf{z}^*)} \ \chi_{vt;pTM_T;T_p'}^{*}(\mathbf{z}) \ K^\dagger K \ \chi_{vt;pTM_T;T_p}(\mathbf{z})$$

or, with $K^\dagger = K$,

$$\left\langle \Psi(vt; pTM_T; T_p') \middle| \Psi(vt; pTM_T; T_p) \right\rangle = \left(K^2(vt; pT) \right)_{T_p'T_p}.$$

Now K^2 is an overlap matrix with submatrices, for fixed p, T, which are off-diagonal in T_p. Since K commutes with the $SU(2) \times U(1)$ generators, K^2 is diagonal in p and T and independent of M_T and therefore factors into submatrices for each p, T. (The dependence on v, t will henceforth not be explicitly indicated).

Even in this more challenging case the determination of the K^2 matrix through the relation, $\Gamma(\mathbf{A})K^2 = K^2 \mathbf{z}$, will be greatly simplified through the introduction of the auxiliary operator, $\Lambda_{op.}$, with the property

$$[\Lambda_{op.}, \mathbf{z}] = \Gamma(\mathbf{A}) = -i[\mathbf{z} \times \not{v}] - (\frac{1}{2}v - \Omega)\mathbf{z} + \frac{1}{2}(\mathbf{z} \cdot \mathbf{z})\nabla - \mathbf{z}(\mathbf{z} \cdot \nabla)$$

This commutator equation is again easily solved to yield

$$\Lambda_{op.} = i\not{v} \cdot [\mathbf{z} \times \nabla] - (\frac{v}{2} - \Omega)(\mathbf{z} \cdot \nabla) + \frac{1}{4}(\mathbf{z} \cdot \mathbf{z})\nabla^2 - \frac{1}{2}(\mathbf{z} \cdot \nabla)(\mathbf{z} \cdot \nabla) + \frac{1}{2}(\mathbf{z} \cdot \nabla)$$

Note that the first term is simply

$$i\not{v} \cdot [\mathbf{z} \times \nabla] = -(\not{v} \cdot \mathbf{T}^{coll.})$$

with eigenvalue

$$-\frac{1}{2}[T(T+1) - t(t+1) - T_p(T_p+1)]$$

The operator $(\mathbf{z} \cdot \nabla)$ measures the degree, p, of the z-space polynomial, $Z^{(p0)}(\mathbf{z})$; ie it counts the number of $J = 0$, $T = 1$-coupled pairs and has the eigenvalue p. The eigenvalue of the operator, $(\mathbf{z} \cdot \mathbf{z})\nabla^2$, can easily be evaluated by direct calculation. Since it is a scalar in isospin space and therefore independent of M_{T_p}, it is sufficient to let it act on the state with $M_{T_p} = T_p$. It is also convenient to express $(\mathbf{z} \cdot \mathbf{z})$ and the z-space Laplacian in terms of spherical coordinates

$$(\mathbf{z} \cdot \mathbf{z}) = (-2z_{+1}z_{-1} + z_0^2); \qquad \nabla^2 = (-2\frac{\partial^2}{\partial z_{+1} \partial z_{-1}} + \frac{\partial^2}{\partial z_0^2})$$

The differentiations give

$$(\mathbf{z} \cdot \mathbf{z})\nabla^2 Z^{(p0)}_{T_p M_{T_p}} = [p(p+1) - T_p(T_p+1)]Z^{(p0)}_{T_p M_{T_p}}$$

so that the eigenvalue of $\Lambda_{op.}$ is

$$\Lambda = -\frac{1}{2}T(T+1) + \frac{1}{4}T_p(T_p+1) + \frac{1}{2}t(t+1) - (\frac{v}{2} - \Omega)p - \frac{1}{4}p^2 + \frac{3}{4}p.$$

The important quantities are the differences of eigenvalues. These are given by

$$\Lambda_{p+1,T'_p,T'} - \Lambda_{p,T_p,T} = \Omega - \frac{v}{2} + \frac{1}{2} - \frac{p}{2} - \frac{1}{2}[T'(T'+1) - T(T+1)]$$

$$+ \frac{1}{4}[T'_p(T'_p+1) - T_p(T_p+1)].$$

The equation for the determination of K^2 is again solved in the form

$$(\Lambda_{op}.\mathbf{z} - \mathbf{z}\Lambda_{op}.)K^2 = K^2\mathbf{z}.$$

One possible method of finding the K^2 matrix elements is obtained by letting this relation act on a specific ket, $|(p-1)[t \times T'_p]T'M'_T\rangle$, on the right and a specific, $\langle p[t \times T_p]TM_T|$, on the left. Since the M_T-dependent Wigner coefficients are the same for the left and right hand side matrix elements, the final relation can be expressed in terms of angular momentum reduced matrix elements of $\mathbf{z}$ and leads to the relation

$$\sum_{T''_p} \left(\Lambda_{pT_pT} - \Lambda_{(p-1)T''_pT'}\right) \langle p[t \times T_p]T\|\mathbf{z}\|(p-1)[t \times T''_p]T'\rangle \left(K^2(p-1,T')\right)_{T''_pT'_p}$$

$$= \sum_{\overline{T}_p} \left(K^2(p,T)\right)_{T_p\overline{T}_p} \langle p[t \times \overline{T}_p]T\|\mathbf{z}\|(p-1)[t \times T'_p]T'\rangle$$

If the K^2 for $(p-1)$ are known, this leads to a system of linear equations for the determination of the K^2 matrix elements for p. We therefore have a recursive procedure for the determination of K^2, starting with $K^2(0,t) = 1$. In the special case, when T'_p is uniquely determined by the quantum numbers $(p-1)$ and T', and T_p is uniquely determined by p and T; ie when both $K^2(p-1,T')$ and $K^2(p,T)$ are 1×1 submatrices, the reduced matrix elements of $\mathbf{z}$ drop out of the equation, and we have the simple recursive relation

$$\frac{\left(K^2(p,T)\right)_{T_pT_p}}{\left(K^2(p-1,T')\right)_{T'_pT'_p}} = \left(\Lambda_{pT_pT} - \Lambda_{(p-1)T'_pT'}\right).$$

This multiplicity-free case is quite common for the states of actual physical interest. In the case with multiplicities, a second more direct method for the determination of the K^2 matrix elements can be attained by taking the scalar product of the z-vector K^2 equation with the ∇ operator

$$\sum_i (\Lambda_{op.} z_i - z_i \Lambda_{op.}) K^2 \nabla_i = K^2 (\mathbf{z} \cdot \nabla)$$

By taking matrix elements between states, $|p[t \times \overline{T}_p]TM_T\rangle$, on the right and, $\langle p[t \times T_p]TM_T|$, on the left, the right hand side of this relation gives us a single K^2 matrix element $\times$ p. By expressing the scalar product, $\sum_i z_i \nabla_i$, on the left in terms of spherical coordinates, $\sum_m z_m \partial/\partial z_m$, and identifying the angular momentum tensor character of the derivative operator: $\partial/\partial z_m = (-1)^{1-m}[\nabla]^1_{-m}$, this single K^2 matrix element is given by

$$p\left(K^2(p,T)\right)_{T_p\overline{T}_p} = \sum_{T'} \sum_{m,M'_T} (-1)^{1-m} \langle TM_T\, 1-m|T'M'_T\rangle \langle T'M'_T\, 1m|TM_T\rangle$$

$$\times \sum_{T''_pT'_p} \left(\Lambda_{pT_pT} - \Lambda_{(p-1)T''_pT'}\right) \langle p[t \times T_p]T\|z\|(p-1)[t \times T''_p]T'\rangle$$

$$\times \left(K^2(p-1,T')\right)_{T''_pT'_p} \langle (p-1)[t \times T'_p]T'\|\nabla\|p[t \times \overline{T}_p]T\rangle$$

By using a $1 \leftrightarrow 3$ interchange in the Wigner coefficient with $-m$, the m-sums can be performed via the orthogonality of the Wigner coefficients. By using the additional symmetry property

$$(-1)^{T-T'+1}\sqrt{\frac{(2T'+1)}{(2T+1)}}\langle(p-1)[t\times T'_p]T'\|\nabla\|p[t\times\overline{T}_p]T\rangle$$

$$= \langle p[t\times\overline{T}_p]T\|z\|(p-1)[t\times T'_p]T'\rangle$$

the final expression can be put into the form

$$(K^2(p,T))_{T_p\overline{T}_p} = \frac{1}{p}\sum_{T'}\sum_{T'_pT''_p}\left(\Lambda_{pT_pT} - \Lambda_{(p-1)T''_pT'}\right)(K^2(p-1,T'))_{T'_pT''_p}$$

$$\times\langle p[t\times T_p]T\|z\|(p-1)[t\times T''_p]T'\rangle\langle p[t\times\overline{T}_p]T\|z\|(p-1)[t\times T'_p]T'\rangle$$

If the K^2 matrix elements for $(p-1)$ are known, this relation gives us a specific single K^2 matrix element for p, ie for states with one more $J=0$, $T=1$-coupled pair. Note, however, that *this* method involves a sum over a larger number of terms than the first method. For both methods it is still necessary to have an expression for the reduced matrix elements of the "collective" rank 1 tensor, z, in a basis in which the intrinsic isospin t is coupled with the "collective" isospin T_p to resultant T. This is again given by standard angular momentum recoupling theory

$$\langle p[t\times T_p]T\|z\|(p-1)[t\times T'_p]T'\rangle = U(tT'_pT1;T'T_p)\langle(p0)T_p\|z\|(p-1,0)T'_p\rangle$$

where the U-coefficient is again a Racah coefficient in unitary form, and the pure z-space reduced matrix element is given in terms of the $SU(3)\supset SO(3)$ reduced Wigner coefficients which were evaluated earlier.

The K^2 overlap matrices can therefore be evaluated for any neutron-proton quasispin irreducible representation, characterized by v, t, (and Ω). The structure of these matrices controls the allowed p, T combinations. Zeros in 1-dimensional

K^2 submatrices or zero eigenvalues in higher-dimensional K^2 submatrices signal the appearance of forbidden states. *Eg*, states with $T = t + p$ with a unique value of $T_p = p$ belong to 1-dimensional K^2 submatrices, given directly by the difference of Λ eigenvalues

$$\frac{\left(K^2(p, T = t + p)\right)_{pp}}{\left(K^2(p-1, T-1)\right)_{p-1,p-1}} = \Lambda_{pp(t+p)} - \Lambda_{p-1,p-1(t+p-1)} = \Omega - \frac{1}{2}v + 1 - t - p$$

so that $(K^2(p, T = t + p))_{pp}$ with $p = \Omega - \frac{1}{2}v - t + 1$ is zero and signals the fact that this has exceeded the maximum possible p-value for this type of state. The maximum T-value is thus given by $T_{max.} = t + p_{max.} = (\Omega - \frac{1}{2}v)$. Iteration of the above leads to

$$\left(K^2(p, T = t + p)\right)_{pp} = \frac{(\Omega - \frac{v}{2} - t)!}{(\Omega - \frac{v}{2} - t - p)!}.$$

To see the structure of the K^2 matrices in greater detail, let us examine the irreducible representations with $t = 1$. The possible T-values for a $j = \frac{11}{2}$ shell

Table of Allowed p, T − values for $\Omega = 6$, $v = 2$, $t = 1$.

n	p		T				
22	10		1				
20	9	0	1	2			
18	8		1^2	2	3		
16	7	0	1	2^2	3	4	
14	6		1^2	2	3^2	4	5
12	5	0	1	2^2	3	4^2	5
10	4		1^2	2	3^2	4	5
8	3	0	1	2^2	3	4	
6	2		1^2	2	3		
4	1	0	1	2			
2	0		1				

and states with $v = 2$, $t = 1$ are shown in the table; where this example has been chosen in order to illustrate by means of the simplest nontrivial case. Irreducible representations with $t = 0$, $(v = 0, 2, ...)$, have states with $T = T_p$. Irreducible representations with $t = \frac{1}{2}$, $(v = 1, 3, ...)$, have states with $T = T_p \pm \frac{1}{2}$ only. Since the allowed T_p-values for a fixed p, (or nucleon number), differ by 2 units, T_p is uniquely determined by p and T in this case also. For irreducible representations with $t = 0$ and $t = \frac{1}{2}$ the K^2 matrices are therefore all 1-dimensional; and $K^{-1}(p, T)$ serves as a simple numerical normalization factor for the states $\left[Z_{T_p}^{(p0)}(\mathbf{A}) \times |v, t\rangle\right]_{TM_T}$, as in our earlier examples. As in these examples, K^2 is determined by successive application of the simple recursive K^2-ratio given by the differences of the Λ eigenvalues.

For irreducible representations with $t = 1$, $(v = 2, 4, ...)$, T_p is no longer uniquely determined by p and T. Note, in particular, that with $p = 2$ the two possible values of T_p, $(= 2 \text{ or } 0)$, lead, with $\mathbf{T} = \mathbf{t} + \mathbf{T_p}$, to the possible T-values: $T = 1, 2, 3$ and $T = 1$, ie to a double occurrence of states with $T = 1$ and hence a 2×2 K^2 submatrix for $p = 2$, $T = 1$. Similarly, for the $\frac{1}{2}$-full shell with n=12, or $p = 5$, (ie 5 $J = 0, T = 1$-pairs added to the $v = 2$ nucleons free of $J = 0, T = 1$ pairs), the possible T_p-values, $T_p = 5; 3; 1$, would lead to $T = 6, 5, 4$; $T = 4, 3, 2$; $T = 2, 1, 0$. Note, however, that $T = 6$ is missing. This state belongs to the category with $T = t + p$ discussed above, and has a zero norm. The maximum possible T value is $\Omega - \frac{1}{2}v = 5$. Except for the states with $T = 0$, the remaining states with $(p - T) =$odd are all 2-fold, with $T_p = T \pm 1$. However, states with $(p - T) =$ even are single for all values of p with a unique T_p value: $T_p = T$. The diagram illustrates a fairly common occurrence. There are many categories of states which lead to 1-dimensional K^2 submatrices even in representations, (such as this example with $t = 1$), where the most general K^2 submatrix has a higher dimension. Thus, besides the category with $T = t + p$, we have a second such 1-dimensional category in this case. The 1-dimensional K^2 submatrices with $p - T =$even can thus be determined recursively, (without summation formulae or the need for the solution of a system of linear equations), from the simple differences of Λ eigenvalues. Starting with the first step, $\left(K^2(p=1, T=1)\right)_{11} = \Lambda_{111} - \Lambda_{001} = (\Omega - \frac{1}{2}v + 1)$, this leads to

$$\left(K^2(p,T)\right)_{T_p=T, T_p=T} = \frac{(\Omega - \frac{1}{2}v + 1)(\Omega - \frac{1}{2}v - 1)!\,\Gamma(\Omega - \frac{1}{2}v + \frac{3}{2})}{(\Omega - \frac{v}{2} - \frac{p}{2} - \frac{T}{2})!\,\Gamma(\Omega - \frac{v}{2} + \frac{3}{2} - \frac{p}{2} + \frac{T}{2})}.$$

In our table the p, T combinations for $n > 12$, (the $\frac{1}{2}$-full shell value), could be obtained by particle-hole conjugation. However, the K^2 matrices automatically enforce the allowed p, T values. Eg, with $T = 2\Omega - p$, (and $p > \Omega - \frac{v}{2}$), ie for states *beyond* the upper boundary of our p, T diagram, the 1-dimensional K^2 matrix element is zero and automatically signals a forbidden state. Similarly, with $T = 2\Omega - v + 1 - p$, for states *on* the upper boundary of our p, T diagram, there are two possible values of $T_p = T \pm 1$, but only a single allowed T-value. This follows at once from the general expression, [Hecht and Elliott, 1985] for the 2×2 K^2-matrices with $t = 1$ and $p - T$=odd. With $T = 2\Omega - v + 1 - p$ these matrices have one zero eigenvalue and therefore only one allowed state with a nonzero overlap matrix, ie with a nonzero existence probability. With $T = 4$, $p = 7$, eg, the K^2-matrix with $T_p = 3$ (1st row/column), $T_p = 5$ (2nd row/column), has the value

$$\frac{44}{3} \begin{pmatrix} 1 & 2\sqrt{5 \cdot 13} \\ 2\sqrt{5 \cdot 13} & 4 \cdot 5 \cdot 13 \end{pmatrix}$$

with eigenvalues, $\lambda = \frac{44}{3}(0, 261)$. The zero eigenvalue automatically reveals that the basis suffers from a problem of overcompleteness.

In the general case, with more than one allowed state for a specific v, t; p, T-combination it is still necessary to "take the square root" of the K^2 matrices in order to convert the states $\left[Z_{T_p}^{(p0)}(\mathbf{A}) \times |v, t)\right]_{T M_T}$ into an orthonormal basis via the K^{-1} operation. Since our method of calculation gives only the K^2 matrix elements this can be achieved by diagonalizing the K^2-submatrices. If U is the unitary matrix which diagonalizes K^2, and λ_i are the eigenvalues of K^2, so that

$$UK^2U^\dagger = \lambda \quad = \begin{pmatrix} \lambda_1 & 0 & \cdots & 0 \\ 0 & \lambda_2 & \cdots & 0 \\ \vdots & \vdots & \ddots & \vdots \\ 0 & 0 & \cdots & \lambda_d \end{pmatrix}$$

we can take the square root of the matrix or its inverse in diagonal form. Note that from its property as an overlap matrix for the states $|\Psi(vt; pTM_T; T_p)\rangle$ we can see that K^2 must be positive semidefinite; ie $\lambda_i \geq 0$.

In the general case, with *all* $\lambda_i > 0$,

$$K = U^\dagger \lambda^{\frac{1}{2}} U; \qquad K^{-1} = U^\dagger \lambda^{-\frac{1}{2}} U;$$

and the orthonormal basis becomes

$$|vt; pTM_T; \nu\rangle = \sum_{T_p} (K^{-1})_{\nu T_p} |\Psi(vt; pTM_T; T_p)\rangle$$

$$= \sum_{i,T_p} U_{\nu i}^\dagger \lambda_i^{-\frac{1}{2}} U_{iT_p} |\Psi(vt; pTM_T; T_p)\rangle$$

Since we can subject the final orthonormal basis, $|...; \nu\rangle$, to a further arbitrary unitary transformation; we could also have chosen

$$|vt; pTM_T; i\rangle = \sum_{T_p} \lambda_i^{-\frac{1}{2}} U_{iT_p} |\Psi(vt; pTM_T; T_p)\rangle.$$

This form is useful in the case when K^2 has zero eigenvalues since this form is still applicable if the λ_i and therefore the orthonormal basis states, $|...; i\rangle$, are restricted to include only the *nonzero* values of λ_i. (Alternately, redundant states could be removed ab initio). However, since the K^2 submatrices are nearly diagonal in many cases of physical interest, the more symmetrical form

$$K_{\nu T_p}^{-1} = \sum_i U_{\nu i}^\dagger \lambda_i^{-\frac{1}{2}} U_{iT_p}$$

is to be preferred. The orthonormal state, $|\ldots;\nu\rangle$, can then be tagged by the label T_p' which is the dominant component of T_p in the orthonormal basis. Then

$$|\Psi(vt;pTM_T;T_p)\rangle = \sum_{T_p'} K_{T_p T_p'}|vt;pTM_T;T_p'\rangle$$

where the ket without the symbol Ψ is the orthonormal basis state; and T_p' in such a state designates the dominant component in the orthonormal basis state (of mixed T_p-character).

In the nuclear seniority scheme for neutron-proton configurations the label, T_p is approximately a good quantum number in the limit, $\Omega \longrightarrow$ large, ie for large j-values where the need for the more powerful vector coherent state techniques is greatest. This is illustrated by the example of our table, (although $\Omega = 6$ is not an extremely large value of Ω). In the $t = 1$ irreducible representation with $\Omega = 6$ there are 10 p,T combinations with 2×2 K^2-submatrices. If these are diagonalized by

$$U = \begin{pmatrix} -\cos\theta & \sin\theta \\ \sin\theta & \cos\theta \end{pmatrix}$$

it is found that the $\cos\theta$-values are ~ 1 for all 10 cases: $\cos\theta$ ranges from a minimum value of .96937 for $n = 18$, $T = 1$ to $\cos\theta = .99919$ for $n = 12$, $T = 4$. The natural orthonormal basis provided by the vector coherent state construction is thus very closely related to the labelling scheme utilizing the physically meaningful label T_p.

Finally, with the establishment of the orthonormal basis through the K^{-1} matrices we can evaluate the angular momentum reduced matrix elements of the pair creation operators

$$\langle vt; p+1T'; T'_{p_\beta} \| \mathbf{A} \| vt; pT; T_{p_\alpha} \rangle$$

$$= \left\langle vt; p+1T'M'_T; T'_{p_\beta} \Big| [\mathbf{A}_1 \times |vt; pT; T_{p_\alpha}\rangle]_{T'M'_T} \right.$$

$$= \left\langle vt; p+1T'M'_T; T'_{p_\beta} \Big| \sum_{T_p} K^{-1}_{T_{p_\alpha}T_p} \Big[\mathbf{A}_1 \times [Z^{(p0)}_{T_p}(\mathbf{A}) \times |v,t\rangle]_T \Big]_{T'M'_T} \right.$$

$$= \sum_{T_p} K^{-1}_{T_{p_\alpha}T_p} \sum_{T'_p} U(tT_pT'1; TT'_p) \langle (p0)T_p; (10)1 \| (p+1,0)T'_p \rangle \sqrt{(p+1)}$$

$$\times \langle vt; p+1T'M'_T; T'_{p_\beta} | [Z^{(p+1,0)}_{T'_p}(\mathbf{A}) \times |v,t\rangle]_{T'M'_T}$$

where we have used the recoupling transformation, (recall that square brackets denote angular momentum coupling, and that a right to left coupling order convention is always used),

$$[\mathbf{A}_1 \times [Z^{(p0)}_{T_p}(\mathbf{A}) \times |v,t\rangle]_T]_{T'M'_T}$$

$$= \sum_{T'_p} U(tT_pT'1; TT'_p) [[\mathbf{A}_1 \times Z^{(p0)}_{T_p}(\mathbf{A})]_{T'_p} \times |v,t\rangle]_{T'M'_T}$$

and

$$[\mathbf{A}_1 \times Z^{(p0)}_{T_p}(\mathbf{A})]_{T'_p} = \sqrt{(p+1)} \langle (p0)T_p; (10)1 \| (p+1,0)T'_p \rangle Z^{(p+1,0)}_{T'_p}(\mathbf{A}).$$

The needed $SU(3) \supset SO(3)$ reduced Wigner coefficients were evaluated earlier.

Finally, with

$$[Z_{T_p'}^{(p+1,0)}(\mathbf{A}) \times |v,t\rangle]_{T'M_T'} = \sum_{T_{T_\beta}'} (K(p+1,T'))_{T_p'T_{p\beta}'} \left| vt; p+1T'M_T'; T_{p\beta}' \right\rangle$$

this leads to

$$\langle vt; p+1, T'; T_{p\beta}' \| \mathbf{A} \| vt; pT; T_{p\alpha} \rangle$$

$$= \sum_{T_p} \sum_{T_p'} (K^{-1}(p,T))_{T_{p\alpha}T_p} (K(p+1,T'))_{T_p'T_{p\beta}'} U(tT_pT'1; TT_p')$$

$$\times \langle (p0)T_p; (10)1 \| (p+1,0)T_p' \rangle \sqrt{(p+1)}.$$

In the special case when the p, T-value for the initial state and the $p+1, T'$-value for the final state both lead to 1-dimensional K-submatrices, (*ie* with unique T_p-values), the ratio of K-values collapses to a single term given by a simple difference of Λ eigenvalues. In that special case:

$$\langle vt; p+1, T'; T_p' \| \mathbf{A} \| vt; pT; T_p \rangle$$

$$= \sqrt{(\Lambda_{p+1,T_p'T'} - \Lambda_{pT_pT})} U(tT_pT'1; TT_p') \langle (p0)T_p; (10)1 \| (p+1,0)T_p' \rangle \sqrt{(p+1)}.$$

This very simple formula also serves as a good approximation formula in the most general case, involving K-submatrices of arbitrary dimension, in the limit of large Ω, where the K-submatrices are approximately diagonal. Even for $\Omega = 6$ this simple formula gives a reasonable approximation. This is illustrated in the following table which gives a few examples comparing this approximation formula for the pair creation operator reduced matrix elements with the exact formula for some of the $v = 2$, $t = 1$ states of the $j = \frac{11}{2}$-shell.

Table of Reduced Matrix Elements for $v = 2$, $t = 1$, $\Omega = 6$.

$$\langle vt; p + 1T'; T'_p \| \mathbf{A} \| vt; pT; T_p \rangle$$

$p+1$	T'	T'_p	p	T	T_p	Exact	Approximate
5	2	1	4	2	2	0.9888	1.0000
5	2	3	4	2	2	1.7385	1.7321
5	2	1	4	3	2	-3.5123	-3.5100
5	2	3	4	3	2	0.3316	0.3703
5	2	1	4	3	4	-0.0365	0
5	2	3	4	3	4	-2.2638	-2.2678

It can be seen that even for $\Omega = 6$ the approximation formula is good to an accuracy of $\sim 1\%$.

The general expression for the reduced matrix elements of the $J = 0$, $T = 1$-pair creation operators can be used to evaluate the matrix elements of a charge dependent pairing interaction

$$H_{pairing} = -\sum_{m_T} g_{m_T} A(m_T) B(m_T).$$

$$\langle vt; pT'M_T; T'_p | H_{pairing} | vt; pTM_T; T_p \rangle =$$

$$-\sum_{m_T} \sum_{T''} \sum_{T''_p} g_{m_T} \langle T''(M_T - m_T) \, 1m_T | TM_T \rangle \langle T''(M_T - m_T) \, 1m_T | T'M_T \rangle$$

$$\times \langle vt; pT; T_p \| \mathbf{A} \| vt; p - 1T'''; T''_p \rangle \langle vt; pT'; T'_p \| \mathbf{A} \| vt; p - 1T'''; T''_p \rangle.$$

In the special case of a charge independent pairing interaction, (g_{m_T} independent of m_T), this interaction becomes diagonal and has eigenvalues given in terms of the Casimir invariants

$$E_{pairing} = -\frac{g}{2}\left[(\Omega - \frac{v}{2})(\Omega - \frac{v}{2} + 3) + t(t+1) - (\Omega - \frac{n}{2})(\Omega - \frac{n}{2} + 3) - T(T+1)\right].$$

This exact result can again be used to test the simple approximation formula for the reduced matrix elements of $\mathbf{A}$. For $n = 12, (p = 5), T = 2$, eg, with $T_p = 1$ and 3, the approximation formula reproduces the diagonal $H_{pairing}$ matrix elements of $-18g$ exactly but would give off-diagonal elements of $0.072g$ compared with the exact value of 0.

In closing this section it may also be useful to point out that we have made very little use of the detailed structure of the full $SO(5)$ group, (or its locally iso-morphic partner $Sp(4)$). Most of our results depended only on a detailed knowl-edge of the much simpler subgroup, $SU(2) \times U(1)$, in particular through its Racah algebra. The quantum numbers, v, t fully specify the irreducible representations of the $SO(5)$ group. If these are given in their standard labelling through the Car-tan subalgebra generated by $H_1 \equiv \frac{1}{2}N_{op.} - \Omega$ and $H_2 \equiv T_0$, we arrive at standard $SO(5)$ irreducible representation labels $(\omega_1, \omega_2) = (\Omega - \frac{1}{2}v, t)$, since the highest weight value of H_1 is given by the maximum possible eigenvalue of $N_{op.} = 4\Omega - v$ and highest possible value of M_T for a state with this nucleon number. The fact that $T_{max.}$ for the irreducible representation has the value $T_{max.} = \Omega - \frac{1}{2}v$ follows at once from symmetry under the Weyl group, but was obtained very simply in our discussion from the structure of the K^2-matrix. To gain standard $Sp(4)$ irreducible representation labels, we should use the Cartan subalgebra in the form $\frac{1}{2}(E_{11} - E_{44}) = \frac{1}{2}(\frac{1}{2}N_{op.} - \Omega + T_0), \frac{1}{2}(E_{22} - E_{33}) = \frac{1}{2}(\frac{1}{2}N_{op.} - \Omega - T_0)$, with highest weight values $[\nu_1, \nu_2] = [\frac{1}{2}(\Omega - \frac{1}{2}v + t), \frac{1}{2}(\Omega - \frac{1}{2}v - t)]$.

3.3. The $Sp(6, R)$ Algebra and Nuclear Collective Motion.

In the past few years the symplectic group Sp(6,R) has emerged through the pioneering contributions of Rosensteel and Rowe[1] as the appropriate dynamical group for a many-body theory of nuclear collective motion. Since $Sp(6, R)$ is also the dynamical group for the three-dimensional harmonic oscillator, with the Elliott $SU(3)$ group as a natural subgroup, it has successfully incorporated core excitations of both quadrupole and monopole type into the shell model foundation of the nuclear collective model and has thus led to the possibility of fully microscopic calculations of nuclear collective phenomena.

The generators for $Sp(6, R)$ were given in chapter 2. in terms of oscillator annihilation and creation operators, a_{si}, $a_{si}^{\dagger}$ for an N-particle system in three dimensions, with $a_{si} = \frac{1}{\sqrt{2}}(x_{si} + \partial/\partial x_{si})$. Now, $s = 1, ..., A$=nucleon number; and $i = 1, 2, 3 \equiv x, y, z$. Since we want to remove the center of mass degree of freedom, we can either replace

$$a_{si} \rightarrow (a_{si} - \frac{1}{A}\sum_{t=1}^{A} a_{ti})$$

or, alternately, replace the A single particle coordinates, s, with $A - 1$ *relative* motion Jacobi coordinates, S. In terms of the latter the $Sp(6, R)$ algebra is given by

$$A_{ij} = A_{ji} = \sum_{S=1}^{A-1} a_{Si}^{\dagger} a_{Sj}^{\dagger}$$

$$B_{ij} = B_{ji} = \sum_{S=1}^{A-1} a_{Si} a_{Sj}$$

$$C_{ij} = \sum_{S=1}^{A-1} a_{Si}^{\dagger} a_{Sj} + \delta_{ij}\frac{(A-1)}{2}$$

where the 6 raising generators, A_{ij}, given here in terms of cartesian components, belong to the symmetric representation, [2], of the $U(3)$ core subgroup, C_{ij}, or

$(\lambda\mu) = (20)$ in Elliott $SU(3)$ notation. In spherical components these contain the $L = 0$ monopole and $L = 2$ quadrupole excitation operators. The 6 hermitian conjugate lowering operators, B_{ij}, belong to the conjugate, $[00 - 2]$, $U(3)$ representation, with $SU(3)$ character $(\lambda\mu) = (02)$. The algebra is characterized by the commutation relations

$$[B_{ij}, A_{kl}] = \delta_{ik}C_{lj} + \delta_{il}C_{kj} + \delta_{jk}C_{li} + \delta_{jl}C_{ki}$$

$$[B_{ij}, C_{kl}] = \delta_{ik}B_{jl} + \delta_{jk}B_{il}$$

$$[C_{ij}, C_{kl}] = \delta_{jk}C_{il} - \delta_{il}C_{kj}$$

and satisfies criteria 1) $-$ 3) of the vector coherent state method. Moreover, any shell model state, ie an A-nucleon state made up of closed shell cores and active nucleons all in the *same* first open major oscillator shell, when coupled to good $U(3)$ symmetry, can serve as a "vacuum" state. Such states are annihilated by the $2\hbar\omega$-lowering operators B_{ij} since none of the valence nucleons can be shifted to lower shells:

$$B_{ij}\|[\sigma]\eta\rangle = 0 \qquad \text{for all} \quad ij, \ \eta.$$

We will use $[\sigma] \equiv [\sigma_1\sigma_2\sigma_3]$ for the $U(3)$ "vacuum" representation; alternately $(\lambda_\sigma\mu_\sigma)N_\sigma$, with $\lambda_\sigma = \sigma_1 - \sigma_2$, $\mu_\sigma = \sigma_2 - \sigma_3$, $N_\sigma = \sigma_1 + \sigma_2 + \sigma_3$. The details of the subgroup labelling, η, will for the most part be unimportant until the last step of any calculation. In the $(0s)^4(0p)^{12}(1s, 0d)^8$ configuration of ^{24}Mg, eg, with $N_\sigma = 28$, not only the dominant ground state $S = 0, T = 0$ component with $(\lambda_\sigma\mu_\sigma) = (84)$, but any other such $SU(3)$ shell model state could serve as a vector "vacuum" or starting state.

The vector coherent state will be built in terms of six complex variables, z_{ij}, which like the A_{ij} are symmetric in the indices: $z_{ij} = z_{ji}$. Using summation convention for repeated (Greek) indices

$$|\mathbf{z}\rangle = e^{\frac{1}{2}z_{\alpha\beta}^* A_{\alpha\beta}}|[\sigma]\eta\rangle = e^{\frac{1}{2}z_{11}^* A_{11} + z_{12}^* A_{12} + \cdots}|[\sigma]\eta\rangle$$

As always, state vectors, $|\psi\rangle$, and operators, O, will be mapped into their z-space realizations:

$$\psi_{[\sigma]\eta}(\mathbf{z}) = \langle \mathbf{z}|\psi\rangle = \langle [\sigma]\eta|e^{\frac{1}{2}z_{\alpha\beta}B_{\alpha\beta}}|\psi\rangle$$

$$\Gamma(O)\psi_{[\sigma]\eta}(\mathbf{z}) = \langle [\sigma]\eta|(O + [\frac{1}{2}z_{\alpha\beta}B_{\alpha\beta}, O] + \cdots)e^{\frac{1}{2}z_{\alpha\beta}B_{\alpha\beta}}|\psi\rangle.$$

Since

$$\Gamma(B_{11}) = 2\frac{\partial}{\partial z_{11}}; \quad \text{while} \quad \Gamma(B_{12}) = \frac{\partial}{\partial z_{12}};$$

it will be convenient to define

$$\nabla_{ij} = (1 + \delta_{ij})\frac{\partial}{\partial z_{ij}}.$$

(Note that $\nabla_{ij} = \nabla_{ji}$). Straightforward application of the commutator algebra gives (in the by now familiar manner)

$$\Gamma(B_{ij}) = \nabla_{ij}$$

$$\Gamma(C_{ij}) = \mathcal{C}_{ij} + z_{i\alpha}\nabla_{\alpha j} = \mathcal{C}_{ij}^{intr.} + C_{ij}^{coll.}$$

$$\Gamma(A_{ij}) = \mathcal{C}_{i\alpha}z_{\alpha j} + \mathcal{C}_{j\alpha}z_{\alpha i} + z_{i\alpha}z_{j\beta}\nabla_{\alpha\beta}$$

The intrinsic $\mathcal{C}_{ij}$, as always, act only on the "vacuum" or starting states with intrinsic quantum numbers $[\sigma]$; whereas the "collective" operators are realized through the six z_{ij} and their derivative partners. We have thus mapped the $Sp(6, R)$ algebra into a direct sum of a six-dimensional oscillator algebra and an intrinsic $U(3)$ algebra. The intrinsic $\mathcal{C}_{ij}$ again commute with the z_{ij} and the ∇_{ij}. Since

$$[\nabla_{ij}, z_{kl}] = \delta_{ik}\delta_{jl} + \delta_{il}\delta_{jk},$$

so that

$$[\nabla_{11}, z_{11}] = 2; \quad \text{while} \quad [\nabla_{12}, z_{12}] = 1,$$

we must, however, normalize the z_{ii} and ∇_{ii} properly. The six z-space oscillator creation and annihilation operators

$$\frac{z_{11}}{\sqrt{2}}, \frac{z_{22}}{\sqrt{2}}, \frac{z_{33}}{\sqrt{2}}, z_{12}, z_{13}, z_{23}; \qquad \frac{\nabla_{11}}{\sqrt{2}}, \frac{\nabla_{22}}{\sqrt{2}}, \frac{\nabla_{33}}{\sqrt{2}}, \nabla_{12}, \nabla_{13}, \nabla_{23}$$

satisfy the standard oscillator commutator algebra and lead to normalized states

$$\left(\frac{z_{11}}{\sqrt{2}}\right)^a \frac{1}{\sqrt{a!}} \left(\frac{z_{22}}{\sqrt{2}}\right)^b \frac{1}{\sqrt{b!}} \left(\frac{z_{33}}{\sqrt{2}}\right)^c \frac{1}{\sqrt{c!}} \frac{(z_{12})^d}{\sqrt{d!}} \frac{(z_{13})^e}{\sqrt{e!}} \frac{(z_{23})^f}{\sqrt{f!}}$$

As in the earlier examples, two tasks lie ahead. The first involves the construction of a properly coupled orthonormal z-space basis. This basis will again be orthonormal with respect to a z-space scalar product with the Bargmann measure. Since our z-space realizations, $\Gamma(A_{ij})$, $\Gamma(B_{ij})$, are not unitary with respect to this measure, the second task again involves the construction of the K operator which effects the transformation to a unitary realization. Since the intrinsic symmetry is now a $U(3)$ symmetry, the six-dimensional oscillator functions must be transformed from the above "cartesian" form to a $Z^{[n_1 n_2 n_3]}(z)$ form, carrying collective $U(3)$ quantum numbers $[n_1 n_2 n_3]$, which can then be coupled to the intrinsic $U(3)$ quantum numbers $[\sigma_1 \sigma_2 \sigma_3]$ of the starting state. The language "intrinsic" and "collective" grew out of this particular application since a nucleus can gain large quadrupole collectivity in two ways: (1) through a large intrinsic quadrupole moment arising from large quantum numbers $[\sigma_1 \sigma_2 \sigma_3]$, in particular large values of $2\lambda_\sigma + \mu_\sigma$ or $\lambda_\sigma + 2\mu_\sigma$; or (2) through the collective core excitations involving the $2\hbar\omega$-oscillator excitation operators which are not only the source of giant quadrupole excitations but enhance the quadrupole collectivity of low-lying states through admixtures of such core excitations.

The double index notation on the six-vector, z_{ij}, is tailor-made for the $U(6) \supset U(3)$ reduction of the six-dimensional oscillator functions. Each z-excitation, *ie* each power of the six-dimensional vector, z, gives us a double excitation of the three-dimensional shell model oscillator; (z_{11} excites two 3-dimensional oscillator quanta in the 1 or x direction, while z_{12} excites one x and one y quantum, etc.) The 6 components of z transform according to the 6-dimensional representation, [2], *ie* the symmetrically coupled representation of two 3-dimensional oscillator quanta. The symmetric polynomial of degree 2 in z carries 4 three-dimensional oscillator quanta, the result of the symmetric coupling of the two [2] excitations:

$$[[2] \times [2]]_{symm.} = [4] + [22].$$

(Note that the [31] representation, present in the general product, $[2] \times [2] = [4] + [31] + [22]$, is missing. It is the antisymmetrically coupled combination of the two identical [2] states). The symmetric polynomial of degree 3 in z carries 6 three-dimensional oscillator quanta, the result of a totally symmetric coupling of three [2] representations, etc. ...

$$[[2] \times [2] \times [2]]_{symm.} = [6] + [42] + [222]$$

$$[[2] \times [2] \times [2] \times [2]]_{symm.} = [8] + [62] + [44] + [422]$$

Or, in general, the symmetric polynomial of degree n carrying $2n$ 3-dimensional oscillator quanta must have a symmetry described by the three-rowed Young tableau, $[n_1 n_2 n_3]$, with $n_1 + n_2 + n_3 = 2n = N$, $n_1 \geq n_2 \geq n_3$; and with n_1, n_2, n_3 all restricted to be *even* numbers only. The construction of the full set of $Z_{\eta n}^{[n_1 n_2 n_3]}$ would be quite complicated. (*Eg* $Z^{[62]}$ has 60 components). For our purposes it will, however, again be sufficient to construct a single very specific component (of "highest" or "lowest" weight); *eg* the one with the maximum possible number of 1 quanta, and subject to this restriction the maximum posssible number of 2 quanta, etc. To avoid confusion between the words "highest" and "lowest", we shall call this the "extremal" state. In terms of the z-dependent "collective" $U(3)$ generators

$$C_{ij}^{coll.} = z_{i\alpha}\nabla_{\alpha j} = (z\cdot\nabla)_{ij}$$

this state satisfies

$$(z\cdot\nabla)_{ii} Z_{extr.}^{[n_1 n_2 n_3]}(z) = n_i Z_{extr.}^{[n_1 n_2 n_3]}(z)$$

$$(z\cdot\nabla)_{ij} Z_{extr.}^{[n_1 n_2 n_3]}(z) = 0; \qquad \text{for } j > i.$$

The last condition, *eg*, makes sure that none of the 2 quanta can be further converted into 1 quanta to raise n_1 and therefore the symmetry of the state. This extremal state can be written in the form

$$Z_{extr.}^{[n_1 n_2 n_3]}(z) = N(x,y,z) z_{11}^x \begin{vmatrix} z_{11} & z_{12} \\ z_{12} & z_{22} \end{vmatrix}^y \begin{vmatrix} z_{11} & z_{12} & z_{13} \\ z_{12} & z_{22} & z_{23} \\ z_{13} & z_{23} & z_{33} \end{vmatrix}^z$$

with $\quad x = \dfrac{1}{2}(n_1 - n_2), \quad y = \dfrac{1}{2}(n_2 - n_3), \quad z = \dfrac{1}{2}n_3.$

The extremal property can be verified at once. *Eg*, with

$$C_{12}^{coll.} = z_{11}\frac{\partial}{\partial z_{12}} + 2z_{12}\frac{\partial}{\partial z_{22}} + z_{13}\frac{\partial}{\partial z_{23}},$$

$$C_{12}^{coll.}(z_{11}z_{22} - z_{12}^2) = 0,$$

$$C_{12}^{coll.}(z_{11}z_{22}z_{33} - z_{11}z_{23}^2 - z_{12}^2 z_{33} + 2z_{12}z_{13}z_{23} - z_{13}^2 z_{22}) = 0,$$

follow at once. (But note that the factor 2 in the operator was needed for the proper cancellation). For the actual calculations, the precise form of the normalization factor $N(x, y, z) \equiv N_{n_1 n_2 n_3}$ is of course vital. (As a small subscript, we also note that $N(x, y, 0)$ can be used to normalize the state $Z_{T_p T_p}^{(p0)}(z)$ of the neutron-proton quasispin algebra if we make the correspondence: $\frac{z_{11}}{\sqrt{2}} \to z_{+1}$, $\frac{z_{22}}{\sqrt{2}} \to z_{-1}$, $z_{12} \to z_0$). The normalization factor, $N(x, y, z)$, was first given by [Quesne, 1981]; but has recently been rederived by [Le Blanc and Rowe, 1987b] by an elegant method applicable to z_{ij}, with $i, j = 1, ..., N$, thus solving the problem of the $\frac{1}{2} N(N+1)$-dimensional oscillator reduced with respect to $U(N)$ for arbitrary N. Since the method, despite its power, is very simple, we shall give it here. It follows from the Capelli operator identity

$$
\begin{vmatrix}
\nabla_{11} & \nabla_{12} & \cdots & \nabla_{1k} \\
\nabla_{21} & \nabla_{22} & \cdots & \nabla_{2k} \\
\vdots & \vdots & \ddots & \vdots \\
\nabla_{k1} & \nabla_{k2} & \cdots & \nabla_{kk}
\end{vmatrix}
\begin{vmatrix}
z_{11} & z_{12} & \cdots & z_{1k} \\
z_{21} & z_{22} & \cdots & z_{2k} \\
\vdots & \vdots & \ddots & \vdots \\
z_{k1} & z_{k2} & \cdots & z_{kk}
\end{vmatrix}
$$

$$
=
\begin{vmatrix}
C_{11} + k + 1 & C_{12} & \cdots & C_{1k} \\
C_{21} & C_{22} + k & \cdots & C_{2k} \\
\vdots & \vdots & \ddots & \vdots \\
C_{k1} & C_{k2} & \cdots & C_{kk} + 2
\end{vmatrix}
$$

where $C_{ij} = (z \cdot \nabla)_{ij}$; and, in the expansion of the determinant of the C_{ik} a column ordering must be used; *ie* the C_{ik} must sit to the right of any $C_{j(k-1)}$, etc. When acting on the extremal state, therefore, the term

$$
\prod_{m=1}^{k} (C_{mm} + k + 2 - m)
$$

is the *only* surviving term from the determinant of the C_{ij}. Therefore

$$
\langle Z_{extr.}^{[n_1 n_2 n_3]} | Z_{extr.}^{[n_1 n_2 n_3]} \rangle = \frac{N_{n_1 n_2 n_3}^2}{N_{n_1 - 2n_2 - 2n_3 - 2}^2} \times
$$

$$\times \left\langle Z_{extr.}^{[n_1-2n_2-2n_3-2]} \right| \begin{vmatrix} \nabla_{11} & \nabla_{12} & \nabla_{13} \\ \nabla_{12} & \nabla_{22} & \nabla_{23} \\ \nabla_{13} & \nabla_{23} & \nabla_{33} \end{vmatrix} \begin{vmatrix} z_{11} & z_{12} & z_{13} \\ z_{12} & z_{22} & z_{23} \\ z_{13} & z_{23} & z_{33} \end{vmatrix} \left| Z_{extr.}^{[n_1-2n_2-2n_3-2]} \right\rangle$$

$$= \frac{N_{n_1n_2n_3}^2}{N_{n_1-2n_2-2n_3-2}^2}(n_1-2+4)(n_2-2+3)(n_3-2+2)$$

$$= \frac{(n_1+2)!!(n_2+1)!!n_3!!}{(n_1-n_3+2)!!(n_2-n_3+1)!!} \frac{N_{n_1n_2n_3}^2}{N_{n_1-n_3n_2-n_30}^2} \left\langle Z_{extr.}^{[n_1-n_3n_2-n_30]} \middle| Z^{[n_1-n_3n_2-n_30]} \right\rangle.$$

Repeating this process $\frac{1}{2}(n_2-n_3)$ times with the 2×2 determinants and $\frac{1}{2}(n_1-n_2)$ times with the 1×1 determinants leads to

$$N_{n_1n_2n_3} = \sqrt{\frac{(n_1-n_2+1)!!(n_1-n_3+2)!!(n_2-n_3+1)!!}{(n_1-n_2)!!(n_1-n_3+1)!!(n_2-n_3)!!(n_1+2)!!(n_2+1)!!n_3!!}}$$

where the double factorial has its usual meaning: $a!! = a(a-2)\cdots2(\text{or }1)$ for a=even(or odd). From the knowledge of the normalization factor of the extremal state, it is now possible to calculate the $U(3)$ or $SU(3)$-reduced matrix elements of z. In general this would also require knowledge of certain $SU(3)$-reduced Wigner coefficients for the coupling $[n_1n_2n_3]\times[2]$, or $(\lambda_n\mu_n)\times(20)$ (in equivalent $SU(3)$ language). By judicious choice of operators, however, it is possible to obtain the reduced matrix elements of z with the use of very simple $SU(3)$-reduced Wigner coefficients which are known to have the value, $+1$. Eg,

$$\left\langle Z_{extr.}^{[(n_1+2)n_2n_3]} \middle| \frac{1}{\sqrt{2}}z_{11} \middle| Z_{extr.}^{[n_1n_2n_3]} \right\rangle = \frac{1}{\sqrt{2}}\frac{N_{n_1n_2n_3}}{N_{(n_1+2)n_2n_3}}$$

$$= (+1)\times\left\langle[(n_1+2)n_2n_3]\|z\|[n_1n_2n_3]\right\rangle,$$

where the factor, $+1$, stands for the full $U(3)$ Wigner coefficient, which for the coupling between the extremal states $[n_1n_2n_3]\to[(n_1+2)n_2n_3]$ describes a 1×1

unitary transformation and has the phase +1 by a generalized Condon-Shortley phase convention. Similarly,

$$\langle Z_{extr.}^{[n_1 n_2 n_3]} | \frac{1}{\sqrt{2}} \nabla_{33} | Z_{extr.}^{[n_1 n_2 (n_3+2)]} \rangle = \sqrt{2} \frac{(n_3+2)}{2} \frac{N_{n_1 n_2 (n_3+2)}}{N_{n_1 n_2 n_3}}$$

$$= (+1) \times \langle [n_1 n_2 n_3] \| \nabla \| [n_1 n_2 (n_3+2)] \rangle$$

where the +1 is again a unique-coupling Wigner coefficient. To gain the required reduced matrix element, we make use of the relation

$$\langle [n_1 n_2 (n_3+2)] \| z \| [n_1 n_2 n_3] \rangle = \sqrt{\frac{dim[n_1 n_2 n_3]}{dim[n_1 n_2 (n_3+2)]}} \langle [n_1 n_2 n_3] \| \nabla \| [n_1 n_2 (n_3+2)] \rangle^*,$$

and the reality of these reduced matrix elements. (Symmetries of $SU(3)$ Wigner coefficients and $SU(3)$-reduced matrix elements of conjugate operators are discussed in the Chapter 5 on $SU(3)$ Wigner coefficients, where care is taken about relative phases). The $SU(3)$ dimension formula gives: $dim[n_1 n_2 n_3] = \frac{1}{2}(n_1 - n_2 + 1)(n_2 - n_3 + 1)(n_1 - n_3 + 2)$. To get the final reduced matrix element of z, we choose

$$\langle Z_{extr.}^{[n_1 (n_2+2) n_3]} | \frac{1}{\sqrt{2}} z_{22} | Z_{extr.}^{[n_1 n_2 n_3]} \rangle = \langle Z_{extr.}^{[n_1 n_2 n_3]} | \frac{1}{\sqrt{2}} \nabla_{22} | Z_{extr.}^{[n_1 (n_2+2) n_3]} \rangle$$

$$= N_{n_1 n_2 n_3} N_{n_1 (n_2+2) n_3} \langle 0 | \nabla_{11}^{\frac{1}{2}(n_1-n_2)} \begin{vmatrix} \nabla_{11} & \nabla_{12} \\ \nabla_{12} & \nabla_{22} \end{vmatrix}^{\frac{1}{2}(n_2-n_3)} \frac{\nabla_{22}}{\sqrt{2}} \begin{vmatrix} \nabla_{11} & \nabla_{12} & \nabla_{13} \\ \nabla_{12} & \nabla_{22} & \nabla_{23} \\ \nabla_{13} & \nabla_{23} & \nabla_{33} \end{vmatrix}^{\frac{1}{2}n_3}$$

$$\times \begin{vmatrix} z_{11} & z_{12} & z_{13} \\ z_{12} & z_{22} & z_{23} \\ z_{13} & z_{23} & z_{33} \end{vmatrix}^{\frac{1}{2}n_3} \begin{vmatrix} z_{11} & z_{12} \\ z_{12} & z_{22} \end{vmatrix}^{\frac{1}{2}(n_2-n_3+2)} z_{11}^{\frac{1}{2}(n_1-n_2-2)} | 0 \rangle$$

Applying the Capelli identity to the 3×3 matrices $\frac{1}{2}n_3$ times, this reduces the above to:

$$N_{n_1n_2n_3}N_{n_1(n_2+2)n_3}\frac{(n_2+3)}{(n_2-n_3+3)}\frac{(n_1+2)!!(n_2+1)!!n_3!!}{(n_1-n_3+2)!!(n_2-n_3+1)!!}$$

$$\times\langle 0|\nabla_{11}^{\frac{1}{2}(n_1-n_2)}\begin{vmatrix}\nabla_{11} & \nabla_{12}\\ \nabla_{12} & \nabla_{22}\end{vmatrix}^{\frac{1}{2}(n_2-n_3)}\frac{\nabla_{22}}{\sqrt{2}}\begin{vmatrix}z_{11} & z_{12}\\ z_{12} & z_{22}\end{vmatrix}^{\frac{1}{2}(n_2-n_3+2)}z_{11}^{\frac{1}{2}(n_1-n_2-2)}|0\rangle$$

$$=\sqrt{2}\frac{(n_2-n_3+2)}{2}\frac{(n_2+3)}{(n_2-n_3+3)}\frac{N_{n_1(n_2+2)n_3}}{N_{n_1n_2n_3}}.$$

This relation can be converted to a $U(3)$-reduced matrix element of z by converting the full matrix element to

$$\langle Z_{extr.}^{[n_1(n_2+2)n_3]}|\frac{1}{\sqrt{2}}z_{22}|Z_{extr.}^{[n_1n_2n_3]}\rangle$$

$$=\langle\cdots\|\cdots\rangle\langle\frac{1}{2}(n_1-n_2)\frac{1}{2}(n_1-n_2)\ 1-1|\frac{1}{2}(n_1-n_2)-1\ \frac{1}{2}(n_1-n_2)-1\rangle$$

$$\times\langle[n_1(n_2+2)n_3]\|z\|[n_1n_2n_3]\rangle,$$

where the first (double-barred) coefficient is the $SU(3)\supset SU(2)$ reduced Wigner coefficient

$$\langle(\lambda_n\mu_n)Y,I=\frac{1}{2}(n_1-n_2);(20)+\frac{2}{3},1\|(\lambda_n-2,\mu_n+2)Y',I'=\frac{1}{2}(n_1-n_2)-1\rangle$$

where Y is the eigenvalue of $\frac{1}{3}(C_{11}+C_{22}-2C_{33})$, and we have used the $SU(3)$ notation $(\lambda_n\mu_n)=((n_1-n_2),(n_2-n_3))$. For the extremal value, $Y'=\frac{1}{3}(n_1+n_2+2-2n_3)$, the values of Y,I and the (20)-Y,I-values, $\frac{2}{3},1$, are uniquely fixed so that this coefficient has the value unity. The ordinary angular momentum,

$SU(2)$, Wigner coefficient has the value $[(n_1 - n_2 - 1)/(n_1 - n_2 + 1)]^{\frac{1}{2}}$. With these results the three $U(3)$ reduced matrix elements of the six-vector $\mathbf{z}$ are given by

$$\langle [(n_1 + 2)n_2 n_3 \| \mathbf{z} \| [n_1 n_2 n_3] \rangle = \sqrt{\frac{(n_1 + 4)(n_1 - n_2 + 2)(n_1 - n_3 + 3)}{2(n_1 - n_2 + 3)(n_1 - n_3 + 4)}}$$

$$\langle [n_1(n_2 + 2)n_3] \| \mathbf{z} \| [n_1 n_2 n_3] \rangle = \sqrt{\frac{(n_2 + 3)(n_1 - n_2)(n_2 - n_3 + 2)}{2(n_1 - n_2 - 1)(n_2 - n_3 + 3)}}$$

$$\langle [n_1 n_2 (n_3 + 2)] \| \mathbf{z} \| [n_1 n_2 n_3] \rangle = \sqrt{\frac{(n_3 + 2)(n_1 - n_3 + 1)(n_2 - n_3)}{2(n_1 - n_3)(n_2 - n_3 - 1)}}.$$

With the establishment of these results for the z-space polynomials, $Z^{[n]}(\mathbf{z})$, the full z-space functions can now be constructed through an $SU(3)$ coupling of the intrinsic and collective states

$$\left[Z^{[n_1 n_2 n_3]}(\mathbf{z}) \times \| [\sigma_1 \sigma_2 \sigma_3] \rangle \right]_{\eta}^{[\omega_1 \omega_2 \omega_3] \varrho} \equiv \left| [[\sigma] \times [\mathbf{n}]][\omega] \varrho; \eta \right\rangle$$

where the square bracket now denotes $U(3)$, (or equivalently $SU(3)$), coupling, again with the convention of a right to left coupling order. The new feature is the multiplicity label, ϱ, which is needed when the $U(3)$ product, $[\sigma] \times [\mathbf{n}]$ contains the resultant $U(3)$ representation, $[\omega]$, with a multiplicity greater than one.

Since these z-space functions furnish an orthonormal basis with respect to the Bargmann measure, transformation to the operator form, involving $Z(\mathbf{A})$, will again require the evaluation of the K matrix elements. Assuming again that the transformation to the unitary z-space realization of the operators can be made with a hermitian K, $(K = K^{\dagger})$, the K^2 matrix elements can be determined from the relation

$$\Gamma(A_{ij})K^2 = K^2 z_{ij}$$

As usual, this will be converted to the form

$$[\Lambda_{op.}, z_{ij}]K^2 = K^2 z_{ij}$$

through the introduction of the auxiliary operator, $\Lambda_{op.}$, with the property

$$[\Lambda_{op.}, z_{ij}] = \Gamma(A_{ij}) = \mathcal{C}_{i\alpha} z_{\alpha j} + \mathcal{C}_{j\alpha} z_{\alpha i} + z_{i\alpha} z_{j\beta} \nabla_{\alpha\beta}$$

This operator equation is solved by

$$\Lambda_{op.} = \mathcal{C}_{\mu\nu} z_{\nu\alpha} \nabla_{\alpha\mu} + \frac{1}{4} z_{\mu\alpha} z_{\nu\beta} \nabla_{\alpha\beta} \nabla_{\mu\nu}$$

$$= \mathcal{C}_{\mu\nu}^{intr.} C_{\nu\mu}^{coll.} + \frac{1}{4} C_{\mu\beta}^{coll.} C_{\beta\mu}^{coll.} - z_{\mu\alpha} \nabla_{\alpha\mu}$$

$$= \frac{1}{2} C_{\mu\nu}^{total} C_{\nu\mu}^{total} - \frac{1}{2} \mathcal{C}_{\mu\nu}^{intr.} \mathcal{C}_{\nu\mu}^{intr.} - \frac{1}{4} C_{\mu\nu}^{coll.} C_{\nu\mu}^{coll.} - z_{\mu\alpha} \nabla_{\alpha\mu},$$

where

$$C_{ij}^{total} = \mathcal{C}_{ij}^{intr.} + C_{ij}^{coll.}; \quad \text{and} \quad C_{ij}^{coll.} = z_{i\alpha} \nabla_{\alpha j}.$$

The eigenvalue of $\Lambda_{op.}$ is thus given by the eigenvalues of the various $U(3)$ Casimir operators, $C_{\mu\nu} C_{\nu\mu}$, and the operator $z_{\mu\alpha} \nabla_{\alpha\mu}$ which counts the total number of oscillator quanta in $Z^{[n]}$ and is thus given by $n_1 + n_2 + n_3$. The $U(3)$ Casimir operators can be put in the form

$$C_{\mu\nu} C_{\nu\mu} = C_{11}^2 + C_{22}^2 + 2C_{21} C_{12} + C_{11} - C_{22} + \dots$$

When acting on an extremal state, with $C_{12} \to 0$, ..., the eigenvalues, h_i, of the surviving C_{ii} give the $U(3)$ Casimir invariants

$$(C_{\mu\nu}C_{\nu\mu})_{eigen} = h_1^2 + h_2^2 + h_3^2 + 2h_1 - 2h_3$$

with

$$h_i(\omega) = \omega_i + \frac{1}{2}(A-1); \quad h_i(\mathbf{n}) = n_i; \quad h_i(\sigma) = \sigma_i + \frac{1}{2}(A-1).$$

Since the intrinsic operators, $\mathcal{C}$, commute with z_{ij}, the intrinsic (purely σ-dependent) part of the above Λ operator can be omitted, so that the eigenvalue becomes

$$\Lambda_{n\omega} = \frac{1}{2}\left[h_1^2(\omega) + h_2^2(\omega) + h_3^2(\omega)\right] + h_1(\omega) - h_3(\omega) - \frac{1}{4}(n_1^2 + n_2^2 + n_3^2)$$

$$-\frac{1}{2}(n_1 - n_3) - (n_1 + n_2 + n_3).$$

The important eigenvalue difference can thus be put in the form

$$\Lambda_{n'\omega'} - \Lambda_{n\omega} = \frac{1}{2}(\Delta\omega_1^2 + \Delta\omega_2^2 + \Delta\omega_3^2)$$

$$+\Delta\omega_1(\omega_1 - \omega_3 - 1) + \Delta\omega_2(\omega_2 - \omega_3 - 2) + \Delta\omega_3(-3) + 2\omega_3 + (A-1) - n_i + i - 1,$$

where $n_i' = n_i + 2$, and $\omega_j' = \omega_j + \Delta\omega_j$. Note that $\sum_j \Delta\omega_j = 2$; but there are two possibilities: $\Delta\omega_j = 2$; or $\Delta\omega_j = 1, \Delta\omega_k = 1$.

In terms of the K the unitary form of the raising operators is given by

$$\gamma(\mathbf{A}) = KzK^{-1}.$$

The inversion of this relation again leads to the transformation

$$\left[Z^{[\mathbf{n}]}(\mathbf{z}) \times \|[\sigma]\rangle\right]_\eta^{[\omega]\varrho} \implies K^{-1}\left[Z^{[\mathbf{n}]}(\mathbf{A}) \times \|[\sigma]\rangle\right]_\eta^{[\omega]\varrho}$$

where the operator K^{-1} which converts

$$\left[Z^{[\mathbf{n}]}(\mathbf{A}) \times \|[\sigma]\rangle\right]_\eta^{[\omega]\varrho} \equiv |\Psi([\sigma]; [\omega]N; [\mathbf{n}]\varrho; \eta)\rangle$$

into an orthonormal basis is again a complicated operator, which is best expressed through its matrix elements. The K matrix will again factor into submatrices, since it is diagonal in the true quantum numbers. These include the $[\sigma]$, which are common to all excitations and specify the $Sp(6, R)$ irreducible representations, and the $[\omega]$ which specify the $U(3)$ subgroup irreducible representations. Since the $U(3)$ representations include the total number of oscillator quanta, the number, N, of *added* oscillator quanta, $N = n_1 + n_2 + n_3$, is also a bona fide quantum number. Finally, since K commutes with the $U(3)$ group, it is independent of the subgroup labels, η. The K matrices therefore factor into submatrices specified by $[\sigma], [\omega]$, and N; but will be off-diagonal in the labels $[n]$ and ϱ, (although there will be many physically interesting states in which $[n]$ is uniquely specified by $[\sigma]$, and $[\omega]$, and the multiplicity label ϱ is not needed, leading to 1-dimensional submatrices). The first step of our method of calculation involves the evaluation of the matrix elements of K^2. In the general case these will be factored into the submatrices

$$\left(K^2([\sigma][\omega]N)\right)_{n\varrho, n'\varrho'},$$

where these again follow from the basic relation

$$(\Lambda z - z\Lambda)K^2 = K^2 z$$

In one method the matrix elements of K^2 are obtained by taking matrix elements of this relation between states, $\left|[[\sigma] \times [n]][\omega]\varrho, \eta\right\rangle$ on the right, and $\left\langle[[\sigma] \times [n']][\omega']\varrho', \eta'\right|$ on the left. Since the K^2 matrix elements are independent of the $U(3)$ subgroup labels, η, this leads to an equation involving only $U(3)$ reduced matrix elements of z

$$\sum_{\bar{n}\ \bar{\varrho}} (\Lambda_{n'\omega'} - \Lambda_{\bar{n}\omega}) \langle [[\sigma] \times [n']][\omega']\varrho'\|z\|[[\sigma] \times [\bar{n}]][\omega]\bar{\varrho}\rangle \left(K^2([\sigma][\omega]N)\right)_{\overline{n}\varrho, n\varrho}$$

$$= \sum_{\overline{n}'\ \bar{\varrho}'} \left(K^2([\sigma][\omega']N+2)\right)_{n'\varrho', \overline{n}'\bar{\varrho}'} \langle [[\sigma] \times [\overline{n}']][\omega]\bar{\varrho}'\|z\|[[\sigma] \times [n]][\omega]\varrho\rangle.$$

If the matrix elements of K^2 for a particular N are known, this leads to a set of linear equations for the K^2-matrix elements for $N + 2$, so that the K^2-matrix elements can be evaluated recursively, starting with the "vacuum" value, $K(N = 0) = 1$. In the special case when *both* $[n]$ and $[n']$ are uniquely specified by $[\omega]$ and $[\omega']$ and the multiplicity labels ϱ and ϱ' are not needed, (so that both K^2 submatrices are 1-dimensional), the single common reduced marix element of z factors out of this equation. In this special case the K^2 ratio is again given simply in terms of the difference of Λ eigenvalues

$$\frac{\left(K^2([\sigma][\omega']N + 2)\right)_{n'n'}}{\left(K^2([\sigma][\omega]N)\right)_{nn}} = \Lambda_{n'\omega'} - \Lambda_{n\omega}.$$

In a second method, one particular matrix element of K^2, for a state with excitation, N, can be obtained from known K^2 matrix elements for $N - 2$, by contracting the above equation in the z with the ∇ operator, and writing it in the form

$$\sum_{\mu,\nu}(\Lambda z_{\mu\nu} - z_{\mu\nu}\Lambda)K^2\nabla_{\nu\mu} = K^2\left(\sum_{\mu,\nu} z_{\mu\nu}\nabla_{\nu\mu}\right).$$

Using the fact that the operator $\sum_{\mu,\nu} z_{\mu\nu}\nabla_{\nu\mu}$ has eigenvalue N, and taking matrix elements of this equation between states, $\left|[[\sigma] \times [\bar{n}]][\omega]\bar{\varrho}, \eta\right\rangle$ on the right, and $\left\langle[[\sigma] \times [n]][\omega]\varrho, \eta\right|$ on the left, this leads to

$$N\left(K^2([\sigma][\omega]N)\right)_{n\varrho,\overline{n\varrho}} =$$

$$\sum_{\mu,\nu}\sum_{[\omega'],\eta'\bar{n}'\bar{\varrho}'}\sum\left\langle[[\sigma] \times [\bar{n}']][\omega']\bar{\varrho}', \eta'\right|\nabla_{\mu\nu}\left|[[\sigma] \times [\bar{n}]][\omega]\bar{\varrho}, \eta\right\rangle\sum_{n'\varrho'}\left(K^2([\sigma][\omega']N - 2)\right)_{n'\varrho',\bar{n}'\bar{\varrho}'}$$

$$\times\left\langle[[\sigma] \times [n]][\omega]\varrho, \eta\right|z_{\mu\nu}\left|[[\sigma] \times [n']][\omega']\varrho', \eta'\right\rangle\left(\Lambda_{n\omega} - \Lambda_{n'\omega'}\right)$$

$$=\sum_{[\omega']}\sum_{n'\varrho'}\sum_{\bar{n}'\bar{\varrho}'}\left\langle[[\sigma] \times [\bar{n}]][\omega]\bar{\varrho}\|z\|[[\sigma] \times [\bar{n}']][\omega']\bar{\varrho}'\right\rangle$$

$$\times\left\langle[[\sigma] \times [n]][\omega]\varrho\|z\|[[\sigma] \times [n']][\omega']\varrho'\right\rangle \times 2\left[\sum_{\eta',\eta_2}\left\langle[\omega']\eta' [2]\eta_2|[\omega]\eta\right\rangle^2\right]$$

$$\times \left(K^2([\sigma][\omega']N-2)\right)_{n'\varrho',\overline{n}'\overline{\varrho}'}(\Lambda_{n\omega}-\Lambda_{n'\omega'}),$$

where we have taken the hermitian conjugate of the (real) matrix element of $\nabla_{\mu\nu}$, have used the fact that

$$\sum_{\mu\nu}(z_{\mu\nu}\nabla_{\mu\nu}) = 2\sum_{\eta_2=1}^{6}(z_{\eta_2}\nabla_{\eta_2}),$$

where z_{η_2} is one of the six (normalized) z-space 6-dimensional oscillator creation operators, and where we have factored the full matrix elements of z_{η_2} into an $SU(3)$-reduced factor and an $SU(3) \supset SU(2)$ Wigner coefficient. Finally, making use of the orthonormality of these $SU(3) \supset SU(2)$ Wigner coefficients to carry out the η sums, we are led to the result

$$\left(K^2([\sigma][\omega]N)\right)_{n\varrho,\overline{n}\varrho} = \frac{2}{N}\sum_{[\omega']}\sum_{n'\varrho'}\sum_{\overline{n}'\overline{\varrho}'}\left(K^2([\sigma][\omega']N-2)\right)_{n'\varrho',\overline{n}'\overline{\varrho}'}(\Lambda_{n\omega}-\Lambda_{n'\omega'})$$

$$\times \langle [[\sigma]\times[n]][\omega]\varrho\|z\|[[\sigma]\times[n']][\omega']\varrho'\rangle\langle [[\sigma]\times[\overline{n}]][\omega]\overline{\varrho}\|z\|[[\sigma]\times[\overline{n}']][\omega']\overline{\varrho}'\rangle.$$

This second method gives one specific K^2 matrix element for excitation, N, in terms of (known) K^2 matrix elements for lower excitation, $N-2$, but at the expense of a sum over a larger number of terms than the first method.

In both methods the $U(3)$ reduced matrix elements of the purely collective operator, z, of $U(3)$ rank [2], need to be known in a $U(3)$-coupled basis in which the intrinsic state, $[\sigma]$, is coupled with a collective state, $[n]$, to resultant $U(3)$ representation, $[\omega]$. This is again given by a straightforward recoupling transformation

$$\langle [[\sigma]\times[n']][\omega']\varrho'\|z\|[[\sigma]\times[n]][\omega]\varrho\rangle = U([\sigma][n][\omega'][2];[\omega]\varrho_-;[n']_-\varrho')\langle[n']\|z\|[n]\rangle.$$

The purely collective $U(3)$ reduced matrix elements of the six-vector z were given earlier. The $U(3)$ recoupling (Racah) coefficient is equivalent to an $SU(3)$

U-coefficient and could therefore also be written in $SU(3)$ notation as

$$U\big((\lambda_\sigma\mu_\sigma)(\lambda_n\mu_n)(\lambda'_w\mu'_w)(20); (\lambda_w\mu_w)\varrho_-; (\lambda'_n\mu'_n)_-\varrho'\big).$$

Where not needed, multiplicity labels are replaced by a _. Such $SU(3)$ U-coefficients are readily available through the computer code of Akiyama and Draayer[2].

Although the need for multiplicity labels does arise frequently in practice and the occurrence of multi-dimensional K^2 submatrices is quite common, many of the physically most relevant states are states with 1-dimensional K^2 matrices. This is illustrated by the "ground state symplectic band" for ^{24}Mg built on the dominant $SU(3)$ shell model component for the $(0s)^4(0p)^{12}(1s,0d)^8$ configuration, with $(\lambda_\sigma\mu_\sigma) = (84)$, $N_\sigma = 28$, or $[\sigma_1\sigma_2\sigma_3] = [16,8,4]$. The $SU(3)$ components, $(\lambda_w\mu_w) = (\omega_1 - \omega_2, \omega_2 - \omega_3)$, are shown for the first three excitations of this symplectic band, $(Sp(6,R)$ representation). States of $2\hbar\omega$ excitation are given by a single action of the raising operator, **A**, and therefore all have $[\mathbf{n}] = [2]$. States of $4\hbar\omega$ excitation do not require multiplicity labels since the

$6\hbar\omega :$ $(14,4)\ (13,3)\ (12,5)\ (2,10)\ (49)\ (38)$

$(10,6)^2\ (12,2)^2\ (68)^2\ (46)^2\ (10,0)^2\ (62)^2\ (87)^2\ (11,1)^2\ (54)^2\ (11,4)^2\ (57)^2\ (81)^2$

$(95)^3\ (10,3)^3\ (76)^3\ (65)^3\ (92)^3\ (73)^3$ and $(84)^5$

$4\hbar\omega :$ $(12,4)\ (10,5)\ (11,3)\ (91)\ (80)\ (56)\ (48)\ (72)\ (67)$

$(10,2)^2\ (86)^2\ (83)^2\ (94)^2\ (75)^2\ (64)^2$

$2\hbar\omega :$ $(10,4)\ (66)\ (82)\ (93)\ (85)\ (74)$

$0\hbar\omega$ "vacuum" state : (84)

collective representations, $[\mathbf{n}] = [4], [22]$, or $(\lambda_n\mu_n) = (40), (02)$, with one zero $SU(3)$ quantum number lead only to multiplicity-free $SU(3)$ couplings. The six representations with double occurrences, such as $(\lambda_\sigma \mu_\sigma) = (10, 2)$, can, however, be built with both $(\lambda_n \mu_n) = (40)$ and (02) and thus lead to 2×2 K^2-submatrices. States of $6\hbar\omega$ excitation with $[\text{n}] = [6], [42]$, and $[222]$, and corresponding $(\lambda_n \mu_n) = (60), (22)$, and (00) have a rich structure of possible $SU(3)$ states $(\lambda_\omega \mu_\omega)$. Now, the product $(84) \times (22)$ leads to multiple occurrences of some of those $(\lambda_\omega \mu_\omega)$ for which the coupling of (84) with (22) involves the addition of squares to all three rows of the Young tableau corresponding to $(\lambda_\sigma \mu_\sigma) = (84)$. Simple states such as $(14, 4)$ are built only through the collective excitation $(\lambda_n \mu_n) = (60)$. Double states, such as $(10, 6)^2, ...,$ can be reached via both (60) and (22) and lead to 2×2 K^2 submatrices. Triple states, such as $(9, 5)^3, ...,$ can again be reached by both (60) and (22), but now the coupling $[(84)] \times (22)] \rightarrow (95)$ has a 2-fold multiplicity, so that the 3×3 submatrices are labelled by $(\lambda_n \mu_n) = (60)$; $(\lambda_n \mu_n) = (22)$, $\varrho = 1$ and 2. Finally, the states $(\lambda_\omega \mu_\omega) = (84)$ within the $6\hbar\omega$ excitation space can be reached via collective states $(\lambda_n \mu_n) = (60)$; $(\lambda_n \mu_n) = (22)$ with $\varrho = 1, 2, 3$; and $(\lambda_n \mu_n) = (00)$, leading to a 5×5 K^2-submatrix.

The excitation space shown in our table was used by Rosensteel, Draayer, and Weeks[3] together with a few additional "stretched" $SU(3)$ representations of type $(8 + N, 4)$ with $N = 8, 10, ..., 20$, (corresponding to the largest possible intrinsic prolate deformations in states of $N\hbar\omega$ oscillator excitation), to give an excellent quantitative account of the low-lying $K = 0$ and $K = 2$ rotational bands in ^{24}Mg. In particular, absolute values of the many observed $E2$ transition probabilities were predicted with the bare proton charge, *ie without* effective charges. The observed $E2$-collectivity is built into the ground state rotational bands through small admixtures of the many $SU(3)$ components shown for the first few excitations of the symplectic $(\lambda_\sigma \mu_\sigma) = (84)$-band.

Our table also gives some indication of the nature of the giant $E2$ resonance in ^{24}Mg. Although the $2\hbar\omega$ excitation region can be expected to have strong admixtures of higher symplectic excitations and be fragmented by coupling to other underlying shell model states, the major components of this resonance should arise through the $2\hbar\omega$ excitations of the $Sp(6, R)$ (84)-band, *ie* the six $SU(3)$ representations $(10, 4)$, (66), (82), (93), (85), (74) which lead to nine 2^+ states; two each from the $K = 0$ and 2 bands of $(10, 4)$, (66), and (82), and one each from the $K = 1$ bands of (93) and (85), and one from the $K = 2$ band of (74); (the $K = 0$ band of (74) is a $1^+, 3^+, 5^+, 7^+$ band). E. J. Reske, (see in particular Figs.17 and 18 of Rowe $(1985)^{1)}$), has shown that these 2^+ states of the symplectic (84)-band contain a large fraction of the energy weighted sum rule $E2$ strength and that this strength distribution and spread is predicted to fall approximately in the region of the observed $E2$ excitations.

In order to convert the states $\left[Z^{[n]}(\mathbf{A}) \times |[\sigma]\rangle\right]_{\eta}^{[\omega]\varrho}$ into an orthonormal basis via the K^{-1} operation; it is again necessary to diagonalize K^2 via a unitary matrix, U,

$$UK^2U^\dagger = \lambda$$

so that, again,

$$K = U^\dagger \lambda^{\frac{1}{2}} U; \qquad K^{-1} = U^\dagger \lambda^{-\frac{1}{2}} U.$$

The case of zero eigenvalues is now very rare. (Rowe, Wybourne, and Butler[4] have shown that for $A > 6$, $A=$ nucleon number, there are no zero eigenvalues, λ_i). In addition[5], in deformed nuclei with $A \geq 12$, and especially in heavy deformed nuclei, the K^2 matrices are very nearly diagonal, so that the orthonormal states, $|\cdots\nu\rangle$,

$$||[\sigma][\omega]\eta\, N;\, \nu\rangle = \sum_{n\varrho} \left(K^{-1}([\sigma][\omega]N)\right)_{\nu, n\varrho} \left[Z^{[n]}(\mathbf{A}) \times |[\sigma]\rangle\right]_{\eta}^{[\omega]\varrho}$$

can also be tagged with the labels $[n]\, \varrho$ of the dominant component: $\nu \equiv [n_{dom.}]\, \varrho_{dom.}$.

The derivation of the $SU(3)$-reduced matrix element of $\mathbf{A}$ is an exact parallel of that for the neutron-proton quasispin algebra of sec. 3.2, (if the angular momentum coupling of sec. 3.2 is replaced by $U(3)$-coupling), and leads to

$$\langle [\sigma][\omega']N + 2; [n']\varrho' \| \mathbf{A} \| [\sigma][\omega]N; [n]\varrho \rangle$$

$$= \sum_{\overline{n\varrho}} \sum_{\overline{n'\varrho'}} \left(K^{-1}([\sigma][\omega]N) \right)_{n\varrho,\overline{n\varrho}} \left(K([\sigma][\omega']N + 2) \right)_{\overline{n'\varrho'},n'\varrho'}$$

$$\times U([\sigma][\overline{n}][\omega'][2]; [\omega]\overline{\varrho}_-; [\overline{n'}]_-\overline{\varrho'}) \langle [\overline{n'}] \| \mathbf{z} \| [\overline{n}] \rangle,$$

where the $[n]\varrho$; $[n']\varrho'$, (without bars), now label the orthonormal states through the dominant components contained in them. In the special case when the K and K^{-1} above are both 1-dimensional submatrices, (with $[n]$ uniquely determined by $[\sigma]$ and $[\omega]$, and $\varrho = 1$ only; ie multiplicity label not needed), the above collapses to the simple formula

$$\langle [\sigma][\omega']N + 2; [n']\varrho' \| \mathbf{A} \| [\sigma][\omega]N; [n]\varrho \rangle =$$

$$\sqrt{(\Lambda_{n'\omega'} - \Lambda_{n\omega})} U([\sigma][n][\omega'][2]; [\omega]\varrho_-; [n']_-\varrho') \langle [n'] \| \mathbf{z} \| [n] \rangle,$$

(with $\varrho=1$; $\varrho'=1$ only). In the limit of large nucleon number and large quantum numbers $[\sigma]$, specifically[5] in the limit, $[\frac{2}{3}(\sigma_1 + \sigma_2 + \sigma_3) + (A - 1)] \rightarrow$ large, the K-submatrices are nearly diagonal to an extremely good approximation; and this simple formula again serves as a good approximation formula in the most general case. In actual applications this approximation formula is in fact spectacularly good[5]. (The parameter $[\frac{2}{3}(\sigma_1 + \sigma_2 + \sigma_3) + (A - 1)]$ now has the value 41.667 for the ground state band in ^{24}Mg. In the corresponding numerical example discussed in sec. 3.2, the validity of the approximation formula was governed by the parameter Ω. The numerical examples given there were for $\Omega = 6$).

Finally, the $SU(3)$ reduced matrix element of $\mathbf{B}$ follows from hermitian conjugation, (phase factors are carefully discussed in Chapter 5 on $SU(3)$ Wigner coefficients),

$$\langle [\sigma][\omega]N; [n]\varrho \| \mathbf{B} \| [\sigma][\omega']N + 2; [n']\varrho' \rangle$$

$$= \sqrt{\frac{dim(\lambda'_\omega \mu'_\omega)}{dim(\lambda_\omega \mu_\omega)}} (-1)^{\lambda_\omega + \mu_\omega + 2 - \lambda'_\omega - \mu'_\omega} \langle [\sigma][\omega']N + 2; [n']\varrho' \| \mathbf{A} \| [\sigma][\omega]N; [n]\varrho \rangle.$$

References for Sec. 3.3.

1) G. Rosensteel and D. J. Rowe, Phys. Rev. Lett. **38**(1977)10; and Ann. Phys. **126**(1980)343. D. J. Rowe, Reports on Progr. in Phys. **48**(1985)1419.

2) Y. Akiyama and J. P. Draayer, Comp. Phys. Commun. **5**(1973)405, and J. Math. Phys. **14**(1973)1904.

3) G. Rosensteel, J. P. Draayer, and K. J. Weeks, Nucl. Phys. **A419**(1984)1.

4) D. J. Rowe, B. G. Wybourne, and P. H. Butler, J. Phys.A: Math. Gen. **18**(1985)939.

5) D. J. Rowe, J. Math. Phys. **25**(1984)2662; and D. J. Rowe, G. Rosensteel, and R. Carr, J. Phys.A:Math. Gen. **17**(1984)L399; and K. T. Hecht, J. PhysA:Math. Gen. **18**(1985)L1003.

3.4 The $SO(6) \supset U(3)$ Algebra and a Relativistic Quark Model of the Nucleus.

The $SO(6) \supset U(3)$ algebra is another simple algebra with a $U(3)$ subgroup. It has recently been suggested[1] that a fermion pair algebra with this structure may be very useful in the detailed calculations needed to investigate the possible validity of a relativistic quark model of the nucleus proposed by the Bonn group[2].

The study of this particular application begins with Fig.4. This sequence of single particle energy levels has most of the characteristics of the conventional nuclear shell model. However, the energy levels shown are those of a zero-rest mass quark in an MIT-type bag with sharp boundary, (mass term $\rightarrow \infty$ at $r = R$). The Bonn quark model of the nucleus is based on the idea that each color singlet three-quark substructure in the low energy states of a nucleus contains one quark pair coupled to $J = 0$, $T = 0$ in its required color $\bar{3}$, $(SU(3) - (01))$-state. Such a pair is therefore spectroscopically inert, and the J, T-structure of an open shell nucleus is determined by the A quarks not in $J = 0$, $T = 0$-coupled pairs and in particular by N such quarks in the unfilled j-subshell, (rather than by 3N such quarks). A $J = 0$, $T = 0$ pairing interaction is introduced to separate the "nonnucleonic" excitations, of Δ-type, eg, from the nuclear states. Such a pairing interaction has also been shown[3] to be a major component of the quark-quark interaction derived by t'Hooft from the instanton solution of QCD. Both its dominant $J = 0$, $T = 0$-pairing characteristic and its strength are in agreement with the phenomenological pairing interaction introduced in the Bonn quark model of the nucleus.

This model not only gives a natural explanation of the characteristics of the Mayer-Jensen shell model level sequence, but also contains the nuclear radius law, $R = r_0 A^{\frac{1}{3}}$, through the equilibrium condition between the inner pressure from quarks in occupied single particle states and the outer vacuum pressure, B, of MIT bag type. Moreover, B is independent of nucleon number, A, and has essentially the MIT value.

Fig. 4.

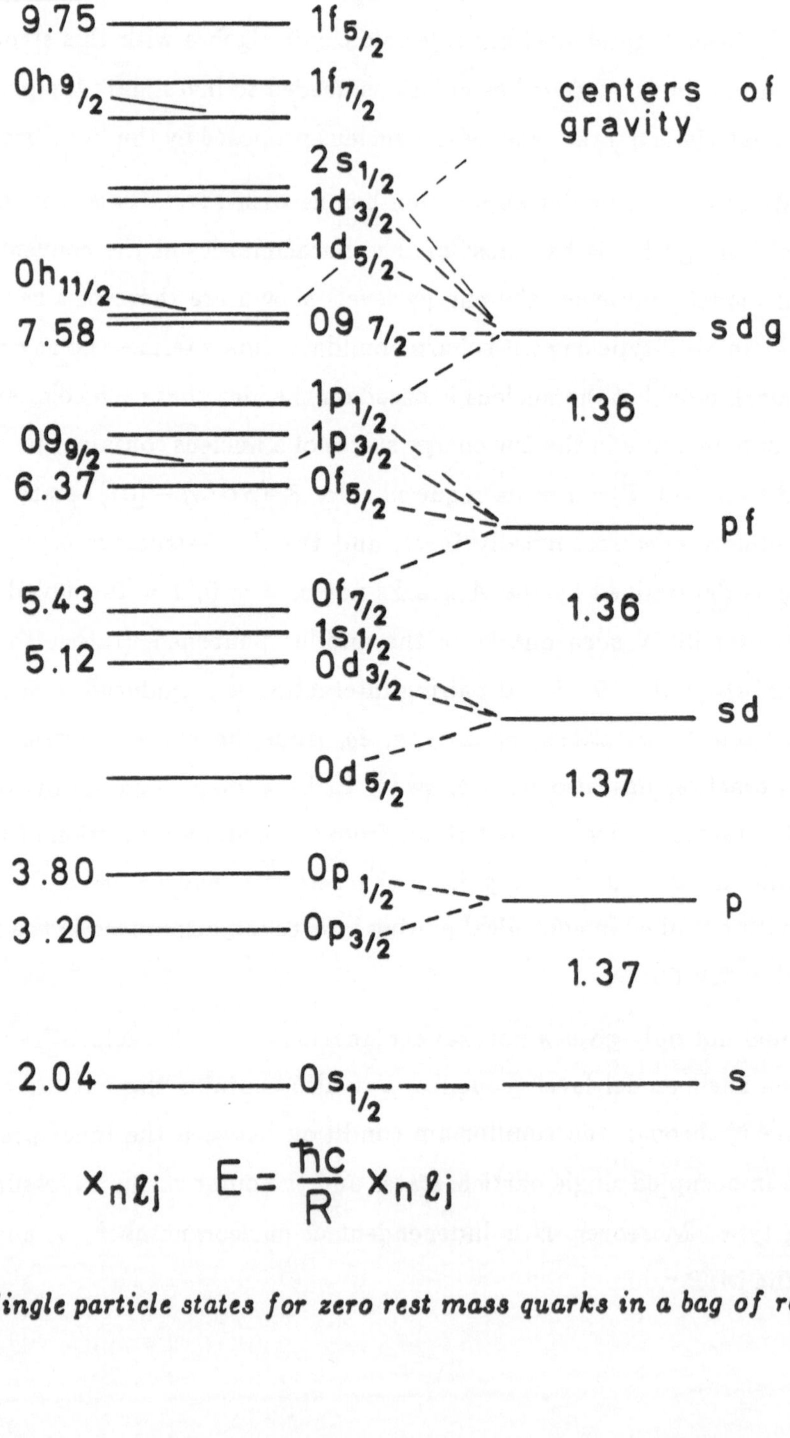

9.75 ———	$1f_{5/2}$	
$0h_{9/2}$ ⟨	$1f_{7/2}$	centers of gravity
	$2s_{1/2}$	
	$1d_{3/2}$	
	$1d_{5/2}$	
$0h_{11/2}$		
7.58	$09\,_{7/2}$ ———	s d g
	$1p_{1/2}$	1.36
$09_{9/2}$	$1p_{3/2}$	
6.37 ———	$0f_{5/2}$	pf
5.43 ———	$0f_{7/2}$	1.36
	$1s_{1/2}$	
5.12 ———	$0d_{3/2}$	sd
	$0d_{5/2}$	1.37
3.80 ———	$0p_{1/2}$	p
3.20 ———	$0p_{3/2}$	1.37
2.04 ———	$0s_{1/2}$	s

$$x_{n\ell j} \qquad E = \frac{\hbar c}{R}\, x_{n\ell j}$$

Single particle states for zero rest mass quarks in a bag of radius R.

In its simplest form, however, this quark model has many deficiencies. In its most naive form one might expect that it would predict 2, 3, and 4-nucleon systems dominated by $0s_{\frac{1}{2}}$ configurations. Nonrelativistic quark models of the few nucleon systems show that strong excitations into the p and higher shells are needed to achieve the clustering into separate three-quark systems. Nonrelativistic quark cluster models predict a $(0s)_{\frac{1}{2}}^6$ probability of at most 1/9 for the two-nucleon system, and a $(0s)_{\frac{1}{2}}^9$ probability of at most 0.004 for the three-nucleon system[4]. A realistic quark model of the nucleus of the Bonn type would therefore require strong configuration mixing, even in heavy nuclei, in order to begin to develop the strong spatial correlations into three-quark clusters which seem to be required for real nuclei. On the other hand, it may be just as good a starting point for a model of the nucleus to consider a system of $3A$ quarks moving in a single bag of nuclear size, rather than a system of A separated three-quark bags of nucleonic size. At realistic nuclear matter densities such three-quark nucleon bags would have to show a large amount of overlap and would constantly be dissolving into each other or more complicated dynamical structures of nonspherical shape, as overlapping nucleon bags move through each other.

In the original Bonn model the quark-quark pairing interaction was treated in terms of the conventional seniority group chain $U(d) \supset SO(d)$, where $d = 2(2j + 1)$ for up and down quarks in the last (open) j-shell. In the mixed configuration calculations needed to develop strong spatial correlations between quarks, an alternate approach based on a complementary $SO(6) \supset U(3)$ group chain may be preferable. This symmetry has the advantage that it applies universally to all j-shells as well as to mixed configurations of several active j-shells.

The $SO(6) \supset U(3)$ Lie algebra is generated by the $J = 0$, $T = 0$, color $\bar{3}$-pair creation operators, their conjugate pair annihilation operators, combined with the color $U(3)$ algebra. In terms of single quark creation operators, $b_{jm,m_t,i}^{\dagger}$, for the jth shell of the quark bag model, with two flavors, $m_t = \pm\frac{1}{2}$, and color indices,

$i = 1, 2, 3$, the $J = 0$, $T = 0$, color $\bar{3}$-pair creation operators, A_{ik}, are

$$A_{ik} = -A_{ki} = \frac{1}{2} \sum_{j,m} \sum_{m_t} (-1)^{j-m+\frac{1}{2}-m_t} \left(b^\dagger_{jm,m_t,i} b^\dagger_{j-m,-m_t,k} - b^\dagger_{jm,m_t,k} b^\dagger_{j-m,-m_t,i} \right).$$

The hermitian conjugate pair annihilation operators are

$$B_{ik} = -B_{ki} = \sum_{j,m} \sum_{m_t} (-1)^{j-m+\frac{1}{2}-m_t} \left(b_{j-m,-m_t,k} b_{jm,m_t,i} - b_{j-m,-m_t,i} b_{jm,m_t,k} \right);$$

and the core group is generated by the $U(3)$ color generators

$$C_{ik} = \sum_{j,m} \sum_{m_t} b^\dagger_{jm,m_t,i} b_{jm,m_t,k} - \frac{\Omega}{2} \delta_{ik},$$

$$\text{with} \qquad \Omega = 2 \sum_j (2j + 1).$$

These 15 operators satisfy the commutation relations of the $SO(6) \supset U(3)$ algebra,

$$[B_{ik}, A_{ab}] = \delta_{ka} C_{bi} + \delta_{ib} C_{ak} - \delta_{ia} C_{bk} - \delta_{kb} C_{ai}$$

$$[B_{ik}, C_{ab}] = \delta_{ka} B_{ib} - \delta_{ia} B_{kb}$$

$$[C_{ik}, C_{ab}] = \delta_{ka} C_{ib} - \delta_{ib} C_{ak}$$

The starting intrinsic states are the states entirely free of $J = 0$, $T = 0$, quark pairs; with color $U(3)$ symmetry $[\sigma] \equiv [\sigma_1 \sigma_2 \sigma_3]$, and color subgroup labels η; where

$$B_{ik} |[\sigma]\eta\rangle = 0 \qquad \text{for all } \eta, \quad ik.$$

In the actual application to the Bonn model, $[\sigma]$ is given by $[\sigma] = [N] \equiv [N00]$ for the purely nuclear states with N spectroscopically active quarks in a state of totally symmetric color symmetry, (characterized by a one-rowed Young tableau), or $[\sigma] = [N\ell\ell]$, with $\ell = 1, 2$ or $=$ a small number, corresponding to ℓ Δ-like $J = \frac{3}{2}, T = \frac{3}{2}$ three-quark systems coupled symmetrically with $(N - \ell)$ additional spectroscopically active quarks. Such ℓ-fold Δ-like excitations can be expected at an excitation energy of $\sim \ell \times (200 - 300 \text{ MeV})$.

To construct the vector coherent state it will be useful to introduce the 3-dimensional vectror, $\mathbf{z}$, where z_1, z_2, z_3 are complex variables. The vector coherent state is then

$$|\mathbf{z}\rangle = e^{z_1^* A_{23} + z_2^* A_{31} + z_3^* A_{12}} |[\sigma]\eta\rangle,$$

where the $U(3)$-invariant scalar product

$$\mathbf{z}^* \cdot \mathbf{A} \equiv \frac{1}{2} \epsilon_{\alpha\beta\gamma} z_\alpha^* A_{\beta\gamma} = \sum_{\eta_0} (z_{\eta_0})^* A_{\eta_0}$$

$$= \sqrt{3} [(\mathbf{z}^*)^{(10)} \times A^{(01)}]^{(00)}$$

has been written in various ways. For most of our purposes the "cartesian" form is the most useful. (Note that ϵ_{ijk} is the totally antisymmetric unit tensor). Sometimes subgroup labels, η_0, expressed in terms of color $SU(2) \times U(1)$ subgroup quantum numbers of the $\bar{3}$ or (01)-representation may also be useful; and the last $SU(3)$-coupled form emphasizes the $SU(3)$-scalar character of $\mathbf{z}^* \cdot \mathbf{A}$. Note that the z_i^* transform according to the $SU(3)$ representation (10) whereas the A_{jk} transform according to the conjugate representation (01). In the last form the square bracket denotes $SU(3)$-coupling. (We note in passing that scalar products of the type $\mathbf{z}^* \cdot \mathbf{A}$ in our earlier applications and examples could also have been written in such a coupled form, involving a coupling to the scalar representation of the core subgroup and very simple core subgroup Wigner coefficients).

As always, state vectors are mapped into their z-space functional representations

$$|\psi\rangle \implies \psi_{[\sigma]\eta}(\mathbf{z}) = \langle \mathbf{z}|\psi\rangle = \langle [\sigma]\eta|e^{\mathbf{z}\cdot\mathbf{B}}|\psi\rangle,$$

where

$$\mathbf{z}\cdot\mathbf{B} = z_1 B_{23} + z_2 B_{31} + z_3 B_{12}.$$

Note that the z_i now transform according to the (01) and the B_{jk} according to the (10) representations of $SU(3)$.

Operators are mapped into their z-space realizations, $\Gamma(O)$, in the by now familiar way. The z-space realizations of the $SO(6)$ generators follow from their commutation relations with $\mathbf{z}\cdot\mathbf{B}$,

$$\Gamma(B_{ij}) = \epsilon_{ijk}\partial_k$$

$$\Gamma(C_{ij}) = \mathcal{C}_{ij} - z_j\partial_i + \delta_{ij}(z_\alpha\partial_\alpha)$$

$$\Gamma(A_{ij}) = \epsilon_{ijk}\left(z_\alpha\mathcal{C}_{\alpha k} - z_k(tr\mathcal{C}) - z_k(z_\alpha\partial_\alpha)\right)$$

(with the shorthand notation, $(\partial/\partial z_k) \equiv \partial_k$, and summation convention for repeated indices. Also, $(tr\mathcal{C}) = \mathcal{C}_{\alpha\alpha}$).

The generators of the $SO(6)$ algebra have therefore been mapped into a direct sum of a 3-dimensional harmonic oscillator algebra and an intrinsic $U(3)$ algebra, $\mathcal{C}_{ij}$. The $\mathcal{C}_{ij}$ act only on the components of the "vacuum" or intrinsic vector $||[\sigma]\eta\rangle$ and commute with the z_k and the ∂_k.

Orthonormal z-space functions can be constructed in terms of z-space polynomials, $Z(\mathbf{z})$. Since the 3-dimensional vector, $\mathbf{z}$, transforms according to the antisymmetric $U(3)$ representation [11], with $SU(3)$-equivalent symmetry (01), a polynomial of degree $p = \ell + m + n$ in the z_i,

$$\frac{z_1^\ell}{\sqrt{\ell!}} \frac{z_2^m}{\sqrt{m!}} \frac{z_3^n}{\sqrt{n!}} = Z_{\ell mn}^{[pp]}(z),$$

transforms according to the $U(3)$ representation, $[pp] \equiv [pp0]$, with $SU(3)$ symmetry $(\lambda_p \mu_p) = (0p)$. Note that in general, (eg for higher symmetries $SO(2d) \supset U(d)$ with $d > 3$), the symmetric coupling of two antisymmetric representations, $[11]$, leads to

$$[[11] \times [11]]_{symm.} = [22] + [1111];$$

but with only three colors, ($d = 3$), it is impossible to make a 4-rowed representation, and $[1111]$ is therefore missing. The totally symmetric coupling of p representations $[11]$ thus leads to the single $U(3)$ representation $[pp]$. The orthonormal z-space basis for the $SO(6) \supset U(3)$ algebra is thus built from the $U(3)$-coupled states

$$\chi_{\eta_\omega}^{[[\sigma] \times [pp]][\omega]}(z) = \left[Z^{[pp]}(z) \times \|[\sigma]\rangle \right]_{\eta_\omega}^{[\omega]},$$

where the square bracket denotes the $U(3)$ coupling $[\sigma] \times [pp] \to [\omega] \equiv [\omega_1 \omega_2 \omega_3]$ or, in equivalent $SU(3)$ language, the coupling $[(\lambda_\sigma \mu_\sigma) \times (0p)] \to (\lambda_\omega \mu_\omega) \equiv (\omega_1 - \omega_2, \omega_2 - \omega_3)$; and η_ω is a convenient set of subgroup labels for $[\omega]$. Note that the coupling $(\lambda_\sigma \mu_\sigma) \times (0p)$ is multiplicity-free, so that no multiplicity labels, ϱ, are needed. Also, p is uniquely determined for a fixed $[\omega]$ and $[\sigma]$. Since the $Z^{(0p)}(z)$ are the complex conjugates of the the 3-dimensional oscillator Bargmann functions $Z^{(p0)}(z^*)$, they satisfy the simple oscillator coupling law

$$\left[Z^{(0p_1)}(z) \times Z^{(0p_2)}(z) \right]^{(\lambda_p \mu_p)} = \delta_{(\lambda_p \mu_p)(0,p_1+p_2)} \sqrt{\frac{(p_1+p_2)!}{p_1! p_2!}} Z^{(0,p_1+p_2)}(z).$$

With $p_1 = 1$ this leads to the standard oscillator reduced matrix element

$$\langle [p+1, p+1] \| z \| [pp] \rangle = \sqrt{(p+1)}.$$

Since these z-space functions again furnish an orthonormal basis with respect to a scalar product with the Bargmann measure, transformation to the operator form, involving $Z^{(Op)}(\mathbf{A})$ will require the evaluation of the K matrix elements. With $K = K^\dagger$, these follow from

$$\gamma(A_{ij}) = K^{-1}\Gamma(A_{ij})K = \left(\gamma(B_{ij})\right)^\dagger = K\epsilon_{ijk}z_k K^{-1}$$

The equation for the determination of K^2

$$\Gamma(A_{ij})K^2 = K^2\epsilon_{ijk}z_k$$

is again simplified through the introduction of the Toronto operator, $\Lambda_{op.}$,

$$[\Lambda_{op.}, z_i] = \frac{1}{2}\epsilon_{ijk}\Gamma(A_{jk}) = z_\alpha \mathcal{C}_{\alpha i} - z_i(tr\mathcal{C}) - z_i(z_\alpha\partial_\alpha).$$

This is satisfied by

$$\Lambda_{op.} = \mathcal{C}_{\alpha\beta}z_\alpha\partial_\beta - (tr\mathcal{C})(z_\alpha\partial_\alpha) - \frac{1}{2}(z_\alpha\partial_\alpha)(z_\beta\partial_\beta) + \frac{1}{2}(z_\alpha\partial_\alpha)$$

$$= -\mathcal{C}_{\alpha\beta}C_{\beta\alpha}^{coll.} - \frac{1}{2}(z_\alpha\partial_\alpha)(z_\beta\partial_\beta) + \frac{1}{2}(z_\alpha\partial_\alpha)$$

$$= -\frac{1}{2}C_{\alpha\beta}^{total}C_{\beta\alpha}^{total} + \frac{1}{2}\mathcal{C}_{\alpha\beta}\mathcal{C}_{\beta\alpha} + \frac{1}{2}C_{\alpha\beta}^{coll.}C_{\beta\alpha}^{coll.} - \frac{1}{2}(z_\alpha\partial_\alpha)(z_\beta\partial_\beta) + \frac{1}{2}(z_\alpha\partial_\alpha),$$

where

$$C_{ij}^{total} = \mathcal{C}_{ij} + C_{ij}^{coll.}; \qquad C_{ij}^{coll.} = -z_j\partial_i + \delta_{ij}(z_\alpha\partial_\alpha).$$

Since the operator $(z_\alpha\partial_\alpha)$ counts the degree of $Z^{(Op)}(\mathbf{z})$, it has the simple eigenvalue $p \ (= \ell + m + n)$. The $U(3)$ Casimir invariants have the eigenvalue, (cf sec 3.3),

$$(C_{\alpha\beta}C_{\beta\alpha})_{eigen} = h_1^2 + h_2^2 + h_3^2 + 2h_1 - 2h_3$$

where the h_i folllow from the eigenvalues of the C_{ii}, (the Cartan subalgebra of $U(3)$), in the extremal state with the maximum possible number of type 1 quarks, and subject to this restriction the maximum possible number of type 2 quarks, etc. From the expression for the generators, $C_{ii} = N_i - \frac{1}{2}\Omega$, where N_i counts the number of quarks of type i, we see that

$$h_i(\omega) = \omega_i - \frac{1}{2}\Omega; \qquad h_i(\sigma) = \sigma_i - \frac{1}{2}\Omega;$$

where the ω_i count the number of squares in the ith row of the Young tableau describing the resultant $U(3)$ symmetry. Since the $\mathcal{C}_{jk}$ commute with z_i, the term $\frac{1}{2}\mathcal{C}_{\alpha\beta}\mathcal{C}_{\beta\alpha}$ does not contribute to $[\Lambda_{op.}, z_i]$ and can be deleted from $\Lambda_{op.}$. The eigenvalue of $\Lambda_{op.}$ then has the value

$$\Lambda = -\frac{1}{2}\left[(\omega_1 - \frac{1}{2}\Omega)^2 + (\omega_2 - \frac{1}{2}\Omega)^2 + (\omega_3 - \frac{1}{2}\Omega)^2 + 2\omega_1 - 2\omega_3\right] + \frac{1}{2}p(p+3).$$

For the most general intrinsic state, $[\sigma_1\sigma_2\sigma_3]$, the $U(3)$ tableau, $[\omega_1\omega_2\omega_3]$, for the final coupled state can in general be obtained by adding the $2p$ squares of the tableau $[pp]$ in the following way: a squares can be added to rows 2 and 3, b squares to rows 1 and 3, and c squares to rows 1 and 2; (see Fig. 5); so that

$$\omega_1 = \sigma_1 + b + c; \qquad \omega_2 = \sigma_2 + a + c; \qquad \omega_3 = \sigma_3 + a + b; \qquad a + b + c = p.$$

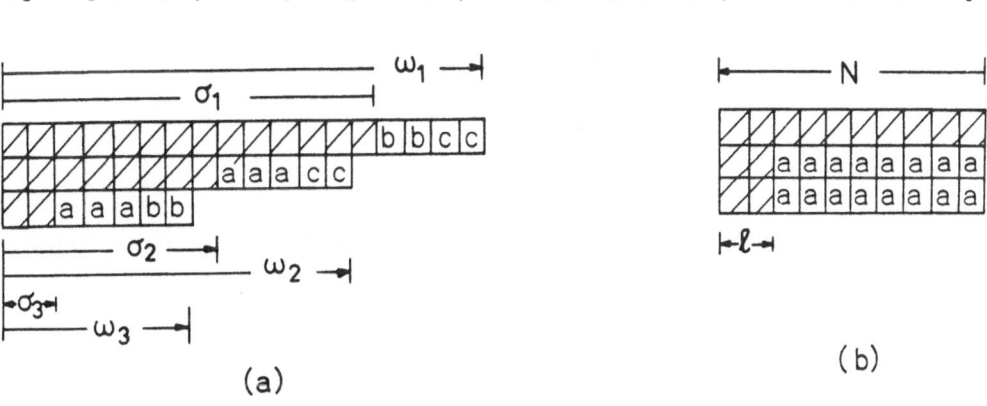

Fig. 5.

The $U(3)$ Symmetries: (a) General case. (b) Color singlet final state of the quark model. The $[\sigma_1\sigma_2\sigma_3]$ are shown by shaded squares. In the quark model of the nucleus involving N active baryons, made up of $3N$ quarks, $\sigma_1 = N$, $\sigma_2 = \sigma_3 = \ell =$ number of Δ-like nonnucleonic excitations.

In terms of the Λ operator the equation for the determination of K^2 again has the form

$$\left(\Lambda_{op.}z_k - z_k\Lambda_{op.}\right)K^2 = K^2 z_k.$$

However, since the coupling $[[\sigma] \times [pp]] \to [\omega]$, or in equivalent $SU(3)$ language $[(\lambda_\sigma\mu_\sigma) \times (0p)] \to (\lambda_\omega\mu_\omega)$, is multiplicity-free, and since p is uniquely determined by $[\omega]$ and $[\sigma]$, all K^2 submatrices, $\left(K^2([\sigma][\omega])\right)$, are 1-dimensional. The matrix elements of the z_i drop out of the above equation; and, with $p' = a' + b' + c' = p + 1 = a + b + c + 1$, the dependence of K^2 on $[\sigma], [\omega]$ can be expressed through the parameters a, b, c.

$$\frac{K^2_{a'b'c'}}{K^2_{abc}} = \Lambda_{a'b'c'} - \Lambda_{abc}$$

In particular

$$\frac{K^2_{a+1,bc}}{K^2_{abc}} = \Lambda_{a+1,bc} - \Lambda_{abc} = \Omega - \sigma_2 - \sigma_3 + 2 - a$$

$$\frac{K^2_{a,b+1,c}}{K^2_{abc}} = \Lambda_{a,b+1,c} - \Lambda_{abc} = \Omega - \sigma_1 - \sigma_3 + 1 - b$$

$$\frac{K^2_{ab,c+1}}{K^2_{abc}} = \Lambda_{ab,c+1} - \Lambda_{abc} = \Omega - \sigma_1 - \sigma_2 - c$$

With $K^2_{000} = 1$, assuming that the intrinsic state, $||\sigma]\rangle$, is normalized, iteration of these relations gives

$$K^2_{abc} = \frac{(\Omega - \sigma_2 - \sigma_3 + 2)!(\Omega - \sigma_1 - \sigma_3 + 1)!(\Omega - \sigma_1 - \sigma_2)!}{(\Omega - \sigma_2 - \sigma_3 + 2 - a)!(\Omega - \sigma_1 - \sigma_3 + 1 - b)!(\Omega - \sigma_1 - \sigma_2 - c)!}.$$

Since K is a simple normalization factor in this symmetry, the orthonormal z-space functions $\chi(z)$ transform into the orthonormal basis in operator form via

$$\chi^{[[\sigma] \times [pp]][\omega]}_{\eta_\omega}(z) = \left[Z^{[pp]}(z) \times ||\sigma]\rangle\right]^{[\omega]}_{\eta_\omega} = K^{-1}\left[Z^{[pp]}(\mathbf{A}) \times ||\sigma]\rangle\right]^{[\omega]}_{\eta_\omega}.$$

In the most general case, $SU(3)$-reduced matrix elements of $\mathbf{A}$ are given by

$$\langle [[\sigma] \times [pp]] [\omega'] \| \mathbf{A} \| [[\sigma] \times [p-1, p-1]] [\omega] \rangle$$

$$= \langle \chi^{[[\sigma] \times [pp]][\omega']}(z) \| \gamma(\mathbf{A}) \| \chi^{[[\sigma] \times [p-1,p-1]][\omega]}(z) \rangle$$

$$= \frac{K_{a'b'c'}}{K_{abc}} \langle \chi^{[[\sigma] \times [pp]][\omega']}(z) \| z \| \chi^{[[\sigma] \times [p-1,p-1]][\omega]}(z) \rangle$$

$$= \frac{K_{a'b'c'}}{K_{abc}} U([\sigma][p-1,p-1][\omega'][11]; [\omega][pp]) \langle [pp] \| z \| [p-1, p-1] \rangle$$

$$= \frac{K_{a'b'c'}}{K_{abc}} U((\lambda_\sigma \mu_\sigma)(0, p-1)(\lambda'_\omega \mu'_\omega)(01); (\lambda_\omega \mu_\omega)(0p)) \sqrt{p}$$

where the needed recoupling coefficient is a simple multiplicity-free Racah coefficient, written in $SU(3)$ form in the last line. This U-coefficient is identically equal to

$$U((\mu_\sigma \lambda_\sigma)(p-1, 0)(\mu'_\omega \lambda'_\omega)(10); (\mu_\omega \lambda_\omega)(p0))$$

$$= U([\sigma_2 - \sigma_3, \sigma_1 - \sigma_2][p-1][\omega'_2 - \omega'_3, \omega'_1 - \omega'_2][1]; [\omega_2 - \omega_3, \omega_1 - \omega_2][p])$$

and can be given in simple analytic form, (see eg eq. A.10 of Le Blanc and Hecht, J. Phys. A:Math. Gen. **20**(1987)...). If the final state $|\omega'|$ is a physically realizable state in the quark model of the nucleus, this must be a color singlet state, with $(\lambda'_\omega \mu'_\omega) = (00)$; and in this case the U-coefficient has the trivial value, $+1$.

The $J = 0, T = 0$-pairing interaction of the Bonn quark model has the simple form

$$H_{pairing} = -g \sum_{\alpha < \beta}^{3} A_{\alpha\beta} B_{\alpha\beta}$$

with eigenvalue

$$E_{pairing}^{[[\sigma] \times [pp]][\omega']} = -g \sum_{[\omega]} \langle [[\sigma] \times [pp]][\omega'] \| \mathbf{A} \| [[\sigma] \times [p-1, p-1]][\omega] \rangle^{2}.$$

For a color singlet $3N$ quark state: $[\omega'] = [NNN]$; and $[\omega]$ has the unique value $[N, N-1, N-1]$. With $[\sigma] = [N\ell\ell]$, $p = a = N - \ell$, (see Fig. 5.),

$$E_{pairing} = -g \frac{K_{p00}^{2}}{K_{p-1,00}^{2}} \times 1 \times p = -g(\Omega - 2\ell + 3 - p)p$$

$$= -g(\Omega - N - \ell + 3)(N - \ell).$$

Although this result can of course also be obtained easily via the $U(\Omega) \supset SO(\Omega)$ group chain[2], the present explicit construction of the state vectors via the $SO(6) \supset U(3)$ symmetry should facilitate the calculations involving more general interactions in a challenging mixed configuration basis of the relativistic quark model of the nucleus.

As a final remark, we note again that we have made very little use of the detailed structure of the full $SO(6)$ group. All of our results again depend only on a detailed knowledge of the simpler subgroup $U(3)$. The irreducible representations of $SO(6)$ are characterized by the quantum numbers $[\sigma_1\sigma_2\sigma_3]$. To get standard $O(6)$ labelling, we need to examine the Cartan subalgebra, $C_{ii} = N_i - \frac{1}{2}\Omega$, with $i = 1, 2, 3$, where N_i counts the number of quarks of type i. For intrinsic states with three-rowed tableaux $[\sigma_1\sigma_2\sigma_3]$: $N_{1_{max}} = \Omega - \sigma_3$; and, with that number of type 1 states filled, $N_{2_{max}} = \Omega - \sigma_2$; and finally $N_{3_{max}} = \Omega - \sigma_1$; leading to *standard* $O(6)$ irreducible representation labelling given by the highest weight eigenvalues of the C_{ii}, *viz* $(\frac{\Omega}{2} - \sigma_3, \frac{\Omega}{2} - \sigma_2, \frac{\Omega}{2} - \sigma_1)$. We note, *eg*, that the pairing interaction above could have been expressed in terms of $SO(6)$ and $U(3)$ Casimir operators, $C^{(C)}$,

$$H_{pairing} = -\frac{g}{2}(C^{(C)}_{SO(6)} - C^{(C)}_{U(3)} + 2C_{\alpha\alpha}),$$

with

$$E_{pairing} = -\frac{g}{2}\left([(\frac{\Omega}{2} - \sigma_3)(\frac{\Omega}{2} - \sigma_3 + 4) + (\frac{\Omega}{2} - \sigma_2)(\frac{\Omega}{2} - \sigma_2 + 2) + (\frac{\Omega}{2} - \sigma_1)^2]\right.$$

$$\left. - [(\omega_1 - \frac{\Omega}{2})^2 + (\omega_2 - \frac{\Omega}{2})^2 + (\omega_3 - \frac{\Omega}{2})^2 + 2(\omega_1 - \omega_3)] + 2(\omega_1 + \omega_2 + \omega_3) - 3\Omega\right).$$

However, the full $O(6)$ representation is usually not needed in our applications which rely solely on a detailed knowledge of the $U(3)$ subalgebra. This is illustrated by the relativistic quark model for a single $j = \frac{3}{2}$ shell, ($\Omega = 8$), and irreducible representation $[\sigma_1\sigma_2\sigma_3] = [311]$, (*ie* $N = 3$, $\ell = 1$), with one Δ-like excitation and two additional "spectroscopically active" quarks. This belongs to the $O(6)$ irreducible representation (331), in *standard* labelling, a 540-dimensional representation. The full set of states of this representation is shown in the table below, which gives the possible n $(\lambda_\omega\mu_\omega)$ combinations, $n=$ particle number, $(\lambda_\omega\mu_\omega) = (\omega_1 - \omega_2, \omega_2 - \omega_3)$. We note that this $O(6)$ representation has two $SO(6)$ branches; a "particle" branch with $(\lambda_\sigma\mu_\sigma) = (20) = (\sigma_1 - \sigma_2, \sigma_2 - \sigma_3)$, $n = \sigma_1 + \sigma_2 + \sigma_3 + 2p$; and a corresponding "hole" branch.

Forbidden states can be picked out via the zeros of K^2. The fact that the state with $n = 15$, $p = 5$, $(\lambda_\omega\mu_\omega) = (25)$, *eg*, is missing follows at once from the K^2 value. For this state $\omega_1 - \sigma_1 = 5$, $\omega_2 - \sigma_2 = 5$, $\omega_3 - \sigma_3 = 0$ so that $abc = 005$; but $(K^2_{005}/K^2_{004}) = \Omega - \sigma_1 - \sigma_2 - 4 = 8 - 4 - 4 = 0$. Similarly, for the state with $n = 17$, $p = 6$, $(\lambda_\omega\mu_\omega) = (15)$, $K^2_{105} = 0$; and for $n = 19$, $p = 7$, $(\lambda_\omega\mu_\omega) = (05)$, $K^2_{205} = 0$; signalling at once that these are forbidden states. For the relativistic quark model of a pure $j = \frac{3}{2}$ configuration the color singlet states with $n = 9$ and

$n = 15$, and $(\lambda_\omega\mu_\omega) = (00)$ are the only $O(6)$ (331) states which would be needed. Most of the members of the $O(6)$-band are nonphysical. However, in a mixed configuration calculation, the $j = \frac{3}{2}$ states with $(\lambda_\omega\mu_\omega) \neq (00)$ may combine with states of matching complementary color, $(\mu_\omega\lambda_\omega)$, from other shells.

The $O(6)$ Irreducible Representation (331). $[\sigma_1\sigma_2\sigma_3] = [311]$.

n	p	$(\lambda_\omega\mu_\omega)$			$(\mu_\omega\lambda_\omega)$		
19	7						(02)
17	6		(04)			(01)	(12)
15	5		(14)	(03)	(00)	(11)	(22)
13	4	(24)	(13)	(02)	(10)	(21)	(32)
11	3	(23)	(12)	(01)	(20)	(31)	(42)
9	2	(22)	(11)	(00)	(30)	(41)	
7	1	(21)	(10)		(40)		
5	0	(20)					

References for Sec. 3.4.

1) K. T. Hecht, Annales Universitatis M. Curie-Sklodowska **41**(1987). Stanislaw Szpikowski 60th birthday issue.

2) K. Bleuler, H. Hofestaedt, S. Merk, and H. R. Petry, Z. Naturf. **38a**(1983)705; H. R. Petry, Lecture Notes in Physics, **197**, (Springer Verlag, Berlin, 1983); and H. Hofestaedt, S. Merk, and H. R. Petry, Z. Phys. **A326**(1987)391.

3) H. R. Petry, H. Hofestaedt, S. Merk, K. Bleuler, H. Bohr, and K. S. Narain, Phys. Lett. **159B**(1985)363.

M. Harvey, Nucl. Phys. **A352**(1981)301, 326; and H. Toki, Y. Suzuki, and K. T. Hecht, Phys. Rev. **C26**(1982)736.

4. Other Applications

In order to give some idea of the range of applicability of the vector coherent state technique we give a brief listing of some other recent applications. The generalization of our discussion of $SU(3)$ of section 3.1. to the general case of the Gelfand chain $SU(n) \supset U(n-1) \supset \cdots$ has been given by [Hecht, Le Blanc and Rowe, 1987b]. Vector coherent state theory has also been used to show that the Wigner-Racah calculus for $SU(n) \supset U(n-1)$ is reduced to an exercise in $U(n-1)$ recoupling, often with known multiplicity-free recoupling coefficients [Le Blanc and Hecht, 1987; Le Blanc,1987].

The fermion pair algebra $SO(2n) \supset U(n)$, for arbitrary n, and its connection with nuclear spectroscopy and boson models for nuclear spectra has been discussed by [Rowe and Carvalho, 1986].

The generalization of the neutron-proton quasispin algebra of section 3.2. to the LST scheme leads to an $SO(8)$ algebra, generated by the 6 $L = 0$-pair creation operators with $S, T = 1, 0$ or $0, 1$, the 6 conjugate $L = 0$-pair annihilation operators, and the $SU(4) + U(1)$ subalgebra of the Wigner supermultiplet + number operator. This $SO(8) \supset U(4)$ algebra has recently been reexamined in terms of vector coherent state theory by [Hecht 1985a]. For a recent application of this algebra as a model in double beta decay theory see [Vogel and Zirnbauer, 1986].

The noncompact group $SO^*(8)$, (which is related to $SO(8)$ much as $Sp(6, R)$ is related to the unitary symplectic group $Sp(6)$), has been proposed by [Le Blanc and Rowe, 1987a] as a beautifully simple mathematical model for $SU(3)$. The algebras $SO^*(2n) \supset U(n)$ and $SO(n, 2) \supset SO(n) + SO(2)$ have also been discussed in a general way by [Le Blanc and Rowe, 1986b] in connection with both mathematical and physical models for $SU(3)$ symmetry.

The S, D fermion pair algebra of [Ginocchio, 1980] also falls into the simplest category of the vector coherent state theory, with core subgroups which, although somewhat more challenging, are simple enough so that the needed Racah coefficients are known or calculable. A fermion pair model, with S and D pairs *only*, the Ginocchio model was originally introduced as a toy model to study the fermionic foundation of the Interacting Boson Model of Arima and Iachello. Recently, it has been proposed as a fermion dynamical symmetry model meriting serious consideration for nuclear spectra throughout the periodic chart, [C.L. Wu et al, 1986], [J.Q. Chen et al, 1986], [R. F. Casten et al, 1986]. In the $SO(8)$ or "i-active" version of this model two of the three possible subgroup chains contain an $SO(n) \supset SO(n-2) \times SO(2)$ chain, with $n = 8$ and 7. Very explicit vector coherent state constructions for these have recently been given by [Hecht, 1987 preprint]. The second "k-active" version of this model leads to a $Sp(6) \supset U(3)$ symmetry, the compact counterpart of the $Sp(6, R) \supset U(3)$ model of section 3.3. This symmetry has been discussed in terms of vector coherent state theory [Hecht, 1985b] in an attempt to elucidate the rotational or $SU(3)$ branch of the Interacting Boson Model of Arima and Iachello.

Very recently the vector coherent state method has been extended through the work of Le Blanc and Rowe to algebras for which the family of raising operators no longer form a family of commuting operators. This has extended the applicability of vector coherent state theory to almost all the classical Lie algebras and their noncompact generalizations. Recent applications include the groups $SO(2n+1) \supset SO(2n) \supset U(n)$, [Rowe, Le Blanc, and Hecht; 1987 preprint], and the exceptional group G_2, [Le Blanc and Rowe; 1987 preprints a,b].

5. The Calculation of SU(3) Wigner Coefficients.

5.1. Introduction

The detailed examples of chapter 3 illustrate how the vector coherent state method and the associated K-matrix construction enable us to give very explicit evaluations of the matrix representations of many higher symmetry algebras. For detailed applications of a particular symmetry, however, it is usually necessary to go further and construct the full Wigner-Racah calculus for that symmetry. Even here, the exploitation of the vector coherent state theory will lead to new simplifications. The calculation of generalized Wigner and Racah coefficients for higher symmetry groups can be achieved essentially from a knowledge of the Wigner and Racah coefficients of the core subgroup and will again make very little use of the detailed structure of the higher symmetry group itself.

To illustrate this powerful simplification for the Wigner-Racah calculus we shall show in this chapter how the vector coherent state construction can be exploited to calculate the simplest $SU(3)$ Wigner coefficients, for the coupling of a general irreducible representation with the fundamental (3-dimensional) representation; ie the $SU(3)$-reduced Wigner coefficients for the coupling $[(\lambda\mu) \times (10)] \rightarrow (\lambda'\mu')$. These coefficients are of course well known, (see eg table 1 of [Vergados 1968]). However, the recognition that such coefficients can be evaluated solely in terms of $SU(2)$ Racah coefficients and a few simple K-matrix elements is new and follows directly from the vector coherent state construction for the $SU(3)$ states. In other more complicated symmetries the evaluation of Wigner coefficients for the coupling of a general irreducible representation with the fundamental representation forms the first step in the evaluation of the Wigner-Racah calculus and can often be used through a buildup process to calculate other simple Wigner coefficients which may be particularly useful in practical applications.

The methods to be illustrated here have been used [Le Blanc and Hecht 1987] to calculate $U(n)$ reduced Wigner coefficients for the $U(n)$ coupling $[f_1...f_n] \times [11...10...0] \equiv [f_1...f_n] \times [1^k]$ for any n and k. Since the notation for general

n becomes somewhat heavy, we shall restrict the discussion to the special case, $n = 3$, where the $SU(2) \times U(1)$ subgroup leads to ordinary angular momentum coupling and recoupling transformations only.

5.2. Preparation.

The explicit construction of orthonormal $SU(3)$ basis states in terms of the action of w raising operators $\mathbf{A}$ was given in section 3.1. The vector coherent state theory leads to the orthonormal basis states

$$|(\lambda\mu)Y(w)IM_I\rangle \equiv \left|(\lambda\mu)[\frac{\lambda}{2} \times \frac{w}{2}]IM_I\right\rangle = \left(K(\lambda\mu)\right)_{w,I}^{-1}\left[Z^{\frac{1}{2}w}(\mathbf{A}) \times \left|\frac{\lambda}{2}\right\rangle\right]_{M_I}^{I},$$

$$\text{with} \qquad Y(w) = \frac{1}{3}(\lambda + 2\mu) - w,$$

and where the square bracket denotes the angular momentum coupling $[\frac{\lambda}{2} \times \frac{w}{2}] \rightarrow IM_I$. The K matrix elements have the values

$$\left(K(\lambda\mu)\right)_{w,I} = \sqrt{\frac{\mu!(\lambda+\mu+1)!}{(\frac{\lambda}{2}+\mu-\frac{w}{2}-I)!(\frac{\lambda}{2}+\mu+1-\frac{w}{2}+I)!}}.$$

In terms of the parameterization,

$$w = p + q, \qquad I = \frac{\lambda}{2} - \frac{p}{2} + \frac{q}{2},$$

these have the even simpler form, (given in section 3.1),

$$\left(K(\lambda\mu)\right)_{p,q} = \sqrt{\frac{\mu!(\lambda+\mu+1)!}{(\mu-q)!(\lambda+\mu+1-p)!}}.$$

This construction led at once to the matrix elements of the group generators which transform according to the 8-dimensional $U(3)$ representation [210], equivalent to the $SU(3)$ representation $(\lambda\mu) = (11)$, using the Elliott notation. For the general discussion of Wigner coefficients other simple tensor operators need to be constructed. In particular, the fundamental tensors of $U(3)$ rank [100], or $(\lambda\mu) = (10)$ in $SU(3)$ notation, and their hermitian conjugates, tensors of $U(3)$ rank $[00-1]$, with $(\lambda\mu) = (01)$, are to be considered in this section. For the explicit construction of such $SU(3)$ tensor operators, including the generators, it will be convenient to introduce a new set of 2 Bargmann 3-vectors,

$$(g_1^a, g_2^a, g_3^a) \qquad \text{with} \quad a = 1, 2,$$

and their hermitian conjugate partners

$$\left(\frac{\partial}{\partial g_1^a}, \frac{\partial}{\partial g_2^a}, \frac{\partial}{\partial g_3^a}\right) \qquad \text{again with} \quad a = 1, 2.$$

Only two Bargmann 3-vectors are needed since the equivalence between $SU(3)$ and $U(3)$ permits us to consider only 2-rowed tableaux with $(\lambda + \mu)$ and μ squares in the first and second rows, (see Fig.2). The subtraction of $SU(3)$-invariant columns of 3 permits us to convert a 3-rowed tableau to a 2-rowed one, eg $[211] \equiv [100]$; or, alternately, the addition of equal numbers of integers converts a $U(3)$ representation with negative integers to one which can be pictured by a 2-rowed tableau, eg $[00-1] \equiv [110]$. (It should perhaps be noted that earlier $U(3)$ constructions using boson operators or Bargmann space variables have often made use of 3×3 such operators or variables. The great simplifications made possible by the seemingly trivial restriction to a set of 2×3 such operators or variables has only recently been fully appreciated through the work of [Le Blanc and Rowe 1986 a,b]). In the 6-dimensional Bargmann space, the operators, g_j^a and their hermitian conjugate, $\partial/\partial g_j^a$, generate a 6-dimensional harmonic oscillator algebra with commutation relations

$$\left[\frac{\partial}{\partial g_k^a}, g_j^b\right] = \delta^{ab}\delta_{kj}.$$

The $U(3)$ generators, E_{kj}, can now be expressed through

$$E_{kj} = \sum_{a=1}^{2} g_k^a \frac{\partial}{\partial g_j^a}; \quad \text{with} \quad A_j \equiv E_{3j} = \sum_{a=1}^{2} g_3^a \frac{\partial}{\partial g_j^a}.$$

On the other hand, the operators

$$F^{ab} = \sum_{j=1}^{3} g_j^a \frac{\partial}{\partial g_j^b}, \qquad \text{with } a, b = 1, 2,$$

generate the group $U(2)$. State vectors constructed from linear combinations of the orthonormal states

$$\frac{(g_1^1)^a}{\sqrt{a!}} \frac{(g_2^1)^b}{\sqrt{b!}} \frac{(g_3^1)^c}{\sqrt{c!}} \frac{(g_1^2)^d}{\sqrt{d!}} \frac{(g_2^2)^e}{\sqrt{e!}} \frac{(g_3^2)^f}{\sqrt{f!}}$$

can thus be classified both by their $SU(3)$, (lower index), and their $U(2)$, (upper index), tensor character. Such double Gelfand polynomials are usually labelled by a double Gelfand pattern with a central row specifying the row structure of the Young tableau common to both lower and upper index symmetries, and an upper and lower pattern specifying the subgroup labels associated with the upper and lower index symmetries. For the general case of $SU(n)$, with states built from $(n-1) \times n$ Bargmann variables, the upper and lower patterns are usually specified by $(n-1)$ rows of tableau row indices corresponding to the subgroup chain $U(n-1) \supset U(n-2) \supset \cdots \supset U(1)$. For the case of $SU(3)$ for which all manipulations can be reduced to ordinary angular momentum coupling, it will be more convenient to use standard angular momentum labels and specify state vectors through the double Gelfand labels

$$\left| \begin{array}{c} \frac{\lambda}{2}m \\ (\lambda\mu) \\ w, IM_I \end{array} \right\rangle,$$

where the central row, written in Elliott $SU(3)$ notation, specifies the number of squares in the 2-rowed tableau, $[\lambda + \mu, \mu]$, which serves both as (lower) $SU(3)$ and (upper) $U(2)$ label. The upper row specifies upper index $SU(2)$ labels in standard angula momentum language, with $m = \frac{1}{2}(n_1 - n_2)$, where n_1 and n_2 are the eigenvalues of F^{11} and F^{22}, respectively. Note that the angular momentum quantum number $\frac{\lambda}{2}$ is redundant, since it is specified by the overhang of the first row of the tableau. It is, however, useful to include this quantum number specifically in the upper label, since we will want to establish a notation which makes it possible to depict angular momentum coupling in both upper and lower pattern angular momenta. The lower row specifies the $U(1) \times SU(2)$ subgroup labels of $SU(3)$ through the standard angular momentum quantum numbers, I, M_I, and the integer, w, which gives the $U(1)$ quantum number $Y = \frac{1}{3}(\lambda + 2\mu) - w$, (or its Elliott counterpart, $\epsilon = -3Y$). Bargmann space functions can then be specified by

$$\left\langle \, g \, \left| \, \begin{matrix} \frac{\lambda}{2}m \\ (\lambda\mu) \\ w, IM_I \end{matrix} \, \right. \right\rangle .$$

Often it will be sufficient to know the extremal state function with the maximum possible number of type 1 quanta in both upper and lower indices, ie with $m = \frac{\lambda}{2}$; $w = 0, I = M_I = \frac{\lambda}{2}$;

$$\left\langle \, g \, \left| \, \begin{matrix} \frac{\lambda}{2}\frac{\lambda}{2} \\ (\lambda\mu) \\ 0, \frac{\lambda}{2}\frac{\lambda}{2} \end{matrix} \, \right. \right\rangle = \frac{1}{N(\lambda\mu)}(g_1^1)^\lambda \begin{vmatrix} g_1^1 & g_2^1 \\ g_1^2 & g_2^2 \end{vmatrix}^\mu ; \quad \text{with} \quad N(\lambda\mu) = \sqrt{\frac{\mu!(\lambda+\mu+1)!}{(\lambda+1)}}.$$

(This normalization coefficient is well known for the general case of $SU(n)$; but note that it could easily be derived by the Capelli operator technique of sec. 3.3).

The Bargmann 3-vectors, $(g_1^a, g_2^a, g_3^a) \equiv g_j^a$, are fundamental tensors of $SU(3)$ rank (10), or $U(3)$ tensors [100], to be denoted by

$$g \begin{pmatrix} \frac{1}{2}m \\ (10) \\ w, IM_I \end{pmatrix} \qquad \text{with } m = +\frac{1}{2}, -\frac{1}{2} \text{ for } a = 1, 2$$

$$Y(w)IM_I = \frac{1}{3}\frac{1}{2} + \frac{1}{2}, \ \frac{1}{3}\frac{1}{2} - \frac{1}{2}, \ -\frac{2}{3}00 \quad \text{for } j = 1, 2, 3$$

Note that the quantum number $Y(w)$ follows uniquely from I in this case and is thus usually omitted. More complicated tensors can be built from these by vector coupling. *Eg*

$$
\left[g \begin{pmatrix} \frac{1}{2} \\ (10) \\ I_1 \end{pmatrix} \times g \begin{pmatrix} \frac{1}{2} \\ (10) \\ I_2 \end{pmatrix} \right]^{00}_{IM_I} \equiv T \begin{pmatrix} 00 \\ (01) \\ w, IM_I \end{pmatrix}
$$

is an $SU(3)$ tensor of rank (01) with $Y(w), I = \frac{2}{3}, 0$ or $-\frac{1}{3}, \frac{1}{2}$; (however, $Y(w)$ again follows uniquely from I so that w is again redundant). The square bracket now denotes angular momentum coupling in *both* the upper and lower angular momenta. Note that the $SU(3)$ character of the resultant tensor is enforced by the upper angular momentum coupling, since the resultant 2-square tableau with an upper $SU(2)$ angular momentum of 0 must be the 1-column tableau $[11]$, $(\equiv (\lambda\mu) = (01))$. Note that the tensor with $I = 0$ can be built only from $I_1 I_2 = \frac{1}{2}\frac{1}{2}$; the tensor with $I = \frac{1}{2}$ only from $I_1 I_2 = \frac{1}{2}0$ or $0\frac{1}{2}$.

Such $SU(3)$ tensors are not yet the famous shift tensors of Biedenharn and Louck, which, when acting on a state of symmetry $(\lambda\mu)$, are supposed to generate a state with one specific $(\lambda'\mu')$; (see *eg* the review by [Louck 1970]). The $g(10)$ tensor, on the other hand, creates a linear combination of states with $SU(3)$ character $(\lambda + 1, \mu)$, $(\lambda - 1, \mu + 1)$, and $(\lambda, \mu - 1)$ when it acts on a state of symmetry $(\lambda\mu)$.

However, the shift characteristics of such operators can be enforced by restrictions on the upper pattern quantum numbers, [Le Blanc and Rowe, 1986a, b]. In particular, the shift character of the operator can be enforced by an angular momentum coupling in the upper pattern quantum numbers. Thus $g_i^{\frac{1}{2}}(10)$ is converted to a shift tensor via

$$
\left[g \begin{pmatrix} \frac{1}{2} \\ (10) \\ i \end{pmatrix} \times \left| \begin{matrix} \frac{\lambda}{2} \\ (\lambda\mu) \\ w, I \end{matrix} \right\rangle \right]^{\frac{\lambda'}{2}m'}_{I'M_I'} ,
$$

where the square bracket again denotes vector coupling in both upper and lower angular momenta. (Note that we could set $m' = \frac{\lambda'}{2}$, its highest weight value, without loss of generality). This converts the $SU(3)$ tensor operator into an operator with specific shift properties and in this sense gives a specific realization of such a shift operator. Now, the resultant state has definite $SU(3)$ character $(\lambda'\mu') = (\lambda+1,\mu)$ or $(\lambda-1,\mu+1)$ which is enforced by the specific choice of the quantum number $\frac{\lambda'}{2} = \frac{1}{2}(\lambda+1)$ or $\frac{1}{2}(\lambda-1)$ of the upper pattern coupling.

However, this leads to an apparent dilemma: This method leads to only *two* shift operations corresponding to the *two* possible values of $\frac{\lambda'}{2} = \frac{1}{2}(\lambda\pm1)$ which result from the addition of one square to the first or second row of the tableau $[\lambda+\mu,\mu]$. In general, however, action with a (10)-tensor can lead to three possible $SU(3)$ states with $(\lambda'\mu') = (\lambda+1,\mu)$, $(\lambda-1,\mu+1)$ *and* $(\lambda,\mu-1)$, corresponding to the addition of one square to rows 1, 2, *or* 3 of the original Young tableau. Since our choice of upper pattern quantum numbers makes 3-rowed tableaux impossible the third possibility seems to be missing. This corresponds to the fact that the angular momentum coupling $[\frac{\lambda}{2} \times \frac{1}{2}] \rightarrow \frac{\lambda'}{2}$ associated with the upper labels excludes the possibility $\frac{\lambda'}{2} = \frac{\lambda}{2}$. However, since the addition of 1 square to row 3 of the original tableau builds an $SU(3)$ representation which is equivalent to one obtained by the removal of 1 square each from both rows 1 and 2, the third type of (10)-shift tensor can be built with the aid of the operator

$$
\left[\frac{\partial}{\partial g} \begin{pmatrix} \frac{1}{2} \\ (01) \\ I_1 \end{pmatrix} \times \frac{\partial}{\partial g} \begin{pmatrix} \frac{1}{2} \\ (01) \\ I_2 \end{pmatrix} \right]_{IM_I}^{00} \equiv T \begin{pmatrix} 0 \\ (10) \\ w, IM_I \end{pmatrix},
$$

and the construction

$$
\left[\left[\frac{\partial}{\partial g} \begin{pmatrix} \frac{1}{2} \\ (01) \\ i_1 \end{pmatrix} \times \frac{\partial}{\partial g} \begin{pmatrix} \frac{1}{2} \\ (01) \\ i_2 \end{pmatrix} \right]_i^0 \middle| \begin{matrix} \frac{\lambda}{2} \\ (\lambda\mu) \\ w, I \end{matrix} \right\rangle \right]_{I'M_I'}^{\frac{\lambda}{2}\frac{\lambda}{2}}.
$$

Note that the operator $(\partial/\partial g_j^\alpha)$ which lowers the degree of a g-space polynomial by 1 unit and thus removes 1 square from a Young tableau, is the hermitian conjugate of the operator g_j^a. Since g_j^a has $U(3)$ character $[100]$, equivalent to the $SU(3)$ rank (10), $\partial/\partial g_j^a$ has $U(3)$ character $[00-1]$, equivalent to $SU(3)$ rank (01). Note that hermitian conjugation converts a general $U(3)$ tensor of rank $[\lambda] = [\lambda_1\lambda_2\lambda_3]$ into one of rank $[\tilde{\lambda}] \equiv [-\lambda_3 - \lambda_2 - \lambda_1]$ and thus converts a $(\lambda\mu) \equiv (\lambda_1 - \lambda_2, \lambda_2 - \lambda_3)$-tensor into a $(\mu\lambda)$-tensor; but does not change the $SU(2)$ or angular momentum quantum numbers, $\frac{\lambda}{2}$, i, associated with upper and lower patterns.

Operators built from both g_j^a and $(\partial/\partial g_j^a)$ will therefore be needed to construct the most general $SU(3)$ tensor. Even for the simple $SU(3)$ coupling $[(\lambda\mu) \times (10)]$ via the fundamental tensor it will sometimes be simpler to act with a $\partial/\partial g$ operator and use a symmetry property which converts the Wigner coefficient for the coupling $[(\lambda'\mu') \times (01)] \rightarrow (\lambda\mu)$ to one for $[(\lambda\mu) \times (10)] \rightarrow (\lambda'\mu')$. For this purpose, however, it will be important to have a consistent phase convention which encompasses both $SU(3)$ and $SU(2)$, and preferably any $SU(n)$. (Since we are interested in generalizations to higher unitary groups, we want to build a consistent phase convention for all n). The hermitian conjugation phase factor should be a generalization of the hermitian conjugation phase factor $\omega(j, m) = j - m$ of $SU(2)$. For an $SU(2)$-tensor

$$(T_{jm})^\dagger = T'_{j,-m}(-1)^{j-m} \equiv T'_{j\tilde{m}}(-1)^{\omega(j,m)}.$$

For $SU(n)$, with $n \geq 3$, the generalization of this phase factor will again depend on both the $SU(n)$ irreducible representation labels and the specific subgroup labels chosen to specify the components of the tensor. For any n

$$\left(T_{[\lambda]_n m}\right)^\dagger = T'_{[\tilde{\lambda}]_n \tilde{m}}(-1)^{\omega([\lambda]_n, m)},$$

where $[\lambda]_n \equiv [\lambda_1\lambda_2\cdots\lambda_n]$ specifies the $U(n)$ representation, and under hermitian conjugation: $[\lambda] \rightarrow [\tilde{\lambda}]$ with $\tilde{\lambda}_j = -\lambda_{n-j+1}$. The label m is a shorthand notation for all the subgroup labels needed to give a complete specification of the components of the tensor. Note that the choice of labels m is often dictated by a particular application. For $SU(3)$-tensors, eg, we might want to use the $U(1) \times SU(2)$ labelling given by w, I, M_I, or possibly a labelling in terms of cartesian tensor components, or in the nuclear physics applications in terms of an $SO(3)$ subgroup, (see Chapter 6). For many purposes, however, we do not need an explicit expression for $w([\lambda], m))$ but only the following fundamental properties, [Hecht, Le Blanc and Rowe, 1987b]:

1) $w([\lambda], m)$ is an integer

2) $\quad w([\lambda], m) + w([\tilde{\lambda}], \tilde{m}) = 2\phi([\lambda])$

3) For a multiplicity-free $U(n)$-coupling $[[\lambda^{(1)}] \times [\lambda^{(2)}]] \to [\lambda^{(3)}]$:

$$w([\lambda^{(1)}], m^{(1)}) + w([\lambda^{(2)}], m^{(2)}) - w([\lambda^{(3)}], m^{(3)}) = \phi([\lambda^{(1)}]) + \phi([\lambda^{(2)}]) - \phi([\lambda^{(3)}]).$$

In 2) and 3) the subgroup-independent phase function, $\phi([\lambda])$, is a generalization of the Condon-Shortley angular momentum phase factor, $j = \frac{1}{2}(\lambda_1 - \lambda_2)$:

$$\phi([\lambda]) = \frac{1}{2} \sum_{j<k}^{n} (\lambda_j - \lambda_k) = \frac{1}{2} \sum_{j=1}^{n} (n + 1 - 2j)\lambda_j.$$

Note that

1) $\phi([\lambda])$ is invariant under hermitian conjugation $\quad \lambda_j \longrightarrow \tilde{\lambda}_j = -\lambda_{n-j+1}$.

2) The quantity $4\phi([\lambda])$ is always an integer.

3) The conjugation combination:

$$\phi([\lambda^{(1)}]) + \phi([\lambda^{(2)}]) - \phi([\lambda^{(3)}])$$

is always an integer.

For $n = 3$:

$$\phi([\lambda + \mu, \mu, 0]) \equiv \phi((\lambda\mu)) = \lambda + \mu.$$

Note also that that $\omega([\lambda]_n, m)$ collapses to $\omega([\lambda]_{n-1}, m')$ if the subgroup labels m contain extremal values for the subgroup labels of $U(n - 1)$. Thus

$$\omega((\lambda\mu)w = 0, I = \frac{\lambda}{2}, M_I) = (\frac{\lambda}{2} - M_I),$$

ie for the extremal state, with $w = 0$, $\omega((\lambda\mu)0\frac{\lambda}{2}M_I)$ collapses to the standard angular momentum hermitian conjugation phase factor.

With these phase relationships symmetry properties for the full (rather than reduced) $U(n)$ Wigner coefficients for *multiplicity-free* couplings $[[\lambda^{(1)}] \times [\lambda^{(2)}]] \rightarrow [\lambda^{(3)}]$ are direct generalizations of the corresponding symmetry properties for the standard angular momentum Wigner coefficients. *Eg*,

I. Symmetry under $1 \leftrightarrow 3$ interchange:

$$\langle [\lambda^{(1)}]m^{(1)}; \ [\lambda^{(2)}]m^{(2)}|[\lambda^{(3)}]m^{(3)}\rangle$$

$$= (-1)^{\phi([\lambda^{(1)}])+\phi([\lambda^{(2)}])-\phi([\lambda^{(3)}])-\omega([\lambda^{(2)}],m^{(2)})}\sqrt{\frac{dim[\lambda^{(3)}]}{dim[\lambda^{(1)}]}}$$

$$\times \langle [\lambda^{(3)}]m^{(3)}; \ [\widetilde{\lambda}^{(2)}]\widetilde{m}^{(2)}|[\lambda^{(1)}]m^{(1)}\rangle.$$

II. Conjugation Symmetry:

$$\langle [\widetilde{\lambda}^{(1)}]\widetilde{m}^{(1)}; \ [\widetilde{\lambda}^{(2)}]\widetilde{m}^{(2)}|[\widetilde{\lambda}^{(3)}]\widetilde{m}^{(3)}\rangle$$

$$= (-1)^{\phi([\lambda^{(1)}])+\phi([\lambda^{(2)}])-\phi([\lambda^{(3)}])}\langle [\lambda^{(1)}]m^{(1)}; \ [\lambda^{(2)}]m^{(2)}|[\lambda^{(3)}]m^{(3)}\rangle.$$

Before proceeding to the actual calculation of the fundamental $SU(3)$ Wigner coefficients we still need to discuss the reduced matrix elements of the tensor operators $g(10)$ and $\partial/\partial g(01)$. It is useful to define two types of reduced matrix elements:

1) an $SU(2)_{upper} \star SU(2)_{lower}$ doubly reduced matrix element, to be denoted by double bars, $\langle \; \| \quad \| \; \rangle$; and

2) an $SU(2)_{upper} \star SU(3)$ doubly reduced matrix element to be denoted by both double bars and double carets, $\langle\langle \; \| \quad \| \; \rangle\rangle$, to emphasize that it is reduced not only with respect to the upper $SU(2)$ group and thus independent of m quantum numbers, but also reduced with respect to $SU(3)$ and thus also independent of lower pattern quantum numbers w, I as well as M_I.

$$
\left\langle \begin{matrix} \frac{\lambda'}{2}\frac{\lambda'}{2} \\ (\lambda'\mu') \\ w', I'M_I' \end{matrix} \; \right| \left[g\left(\begin{matrix} \frac{1}{2} \\ (10) \\ \frac{1}{2} \end{matrix}\right) \times \left|\begin{matrix} \frac{\lambda}{2} \\ (\lambda\mu) \\ w, I \end{matrix}\right\rangle \right] \begin{matrix} \frac{\lambda'}{2}\frac{\lambda'}{2} \\ \\ I'M_I' \end{matrix}
$$

$$
= \left\langle \begin{matrix} \frac{\lambda'}{2} \\ (\lambda'\mu') \\ w', I' \end{matrix} \; \right\| \; g\left(\begin{matrix} \frac{1}{2} \\ (10) \\ \frac{1}{2} \end{matrix}\right) \left\| \begin{matrix} \frac{\lambda}{2} \\ (\lambda\mu) \\ w, I \end{matrix} \right\rangle
$$

$$
= \langle (\lambda\mu)Y(w)I; \; (10)\frac{1}{3}\frac{1}{2} \| (\lambda'\mu')Y'(w')I'\rangle\langle\langle (\lambda'\mu')\|g\|(\lambda\mu)\rangle\rangle,
$$

where the square bracket denotes angular momentum coupling in both the upper and lower angular momenta. The first factor in the last line, with a single double-bar is the $SU(3)$ reduced Wigner coefficient, the quantity which we want to calculate in this section. The above relations follow from the full matrix element

$$
\left\langle \begin{matrix} \frac{\lambda'}{2}\frac{\lambda'}{2} \\ (\lambda'\mu') \\ w', I'M_I' \end{matrix} \; \right| g\left(\begin{matrix} \frac{1}{2}m_u \\ (10) \\ \frac{1}{2}m_\ell \end{matrix}\right) \left| \begin{matrix} \frac{\lambda}{2}(\frac{\lambda}{2}-m_u) \\ (\lambda\mu) \\ w, IM_I \end{matrix} \right\rangle
$$

$$
= \langle \frac{\lambda}{2}(\frac{\lambda'}{2}-m_u) \; \frac{1}{2}m_u | \frac{\lambda'}{2}\frac{\lambda'}{2} \rangle \langle IM_I \; \frac{1}{2}m_\ell | I'M_I' \rangle
$$

$$
\times \left\langle \begin{matrix} \frac{\lambda'}{2} \\ (\lambda'\mu') \\ w', I' \end{matrix} \; \right\| g\left(\begin{matrix} \frac{1}{2} \\ (10) \\ \frac{1}{2} \end{matrix}\right) \left\| \begin{matrix} \frac{\lambda}{2} \\ (\lambda\mu) \\ w, I \end{matrix} \right\rangle
$$

$$= \langle \frac{\lambda}{2}(\frac{\lambda'}{2} - m_u) \frac{1}{2} m_u | \frac{\lambda'}{2} \frac{\lambda'}{2} \rangle \langle IM_I \frac{1}{2} m_\ell | I'M_I' \rangle$$

$$\times \langle (\lambda\mu)Y(w)I; (10)\frac{1}{3}\frac{1}{2} \| (\lambda'\mu')Y'(w')I' \rangle \langle\langle (\lambda'\mu') \| \mathbf{g} \| (\lambda\mu) \rangle\rangle,$$

the orthonormality of the $SU(2)$ Wigner coefficients, and the factorization of the full $SU(3) \supset SU(2) \times U(1)$ Wigner coefficient into an $SU(3)$ reduced Wigner coefficient times the ordinary I-spin $SU(2)$ Wigner coefficient. The $SU(2) \star SU(3)$ doubly reduced matrix element is given by extremal state normalization factors, [Le Blanc and Hecht,1987]

$$\langle\langle (\lambda'\mu') \| \mathbf{g} \| (\lambda\mu) \rangle\rangle = \frac{N(\lambda'\mu')}{N(\lambda\mu)} = \sqrt{\frac{(\lambda' + \mu' + 1)!\mu'!(\lambda + 1)}{(\lambda' + 1)(\lambda + \mu + 1)!\mu!}}.$$

In the calculation of Wigner coefficients it will, however, always factor out of all equations so that its specific value is actually not needed.

In similar fashion

$$\left\langle \begin{matrix} \frac{\lambda}{2}\frac{\lambda}{2} \\ (\lambda\mu) \\ w, IM_I \end{matrix} \left| \left[\frac{\partial}{\partial g} \begin{pmatrix} \frac{1}{2} \\ (01) \\ \frac{1}{2} \end{pmatrix} \left| \begin{matrix} \frac{\lambda'}{2} \\ (\lambda'\mu') \\ w', I' \end{matrix} \right\rangle \right]^{\frac{\lambda}{2}\frac{\lambda}{2}}_{IM_I} \right.$$

$$= \left\langle \begin{matrix} \frac{\lambda}{2} \\ (\lambda\mu) \\ w, I \end{matrix} \left\| \frac{\partial}{\partial g} \begin{pmatrix} \frac{1}{2} \\ (01) \\ \frac{1}{2} \end{pmatrix} \right\| \begin{matrix} \frac{\lambda'}{2} \\ (\lambda'\mu') \\ w', I' \end{matrix} \right\rangle$$

$$= \langle (\lambda'\mu')Y'(w')I'; (01) - \frac{1}{3}\frac{1}{2} \| (\lambda\mu)Y(w)I \rangle \langle\langle (\lambda\mu) \| \partial \| (\lambda'\mu') \rangle\rangle.$$

From hermitian conjugation of the full matrix element for **g** and symmetry properties of the two types of $SU(2)$ Wigner coefficients under $1 \leftrightarrow 3$ interchange, the conjugation phase factors $\omega(\frac{1}{2}, m_u) = \frac{1}{2} - m_u$ and $\omega((10)Y = \frac{1}{3}, \frac{1}{2}m_\ell) \equiv$

$\omega((10)w = 0, \frac{1}{2}m_\ell) = \frac{1}{2} - m_\ell$; the reduced matrix elements of $\mathbf{g}$ and $\partial/\partial\mathbf{g}$ are related by

$$\left\langle \begin{matrix} \frac{\lambda}{2} \\ (\lambda\mu) \\ w, I \end{matrix} \middle\| \frac{\partial}{\partial g} \begin{pmatrix} \frac{1}{2} \\ (01) \\ \frac{1}{2} \end{pmatrix} \middle\| \begin{matrix} \frac{\lambda'}{2} \\ (\lambda'\mu') \\ w', I' \end{matrix} \right\rangle$$

$$= (-1)^{\frac{\lambda'}{2} - \frac{\lambda}{2} + I - I'} \sqrt{\frac{(2I'+1)(\lambda'+1)}{(2I+1)(\lambda+1)}} \left\langle \begin{matrix} \frac{\lambda'}{2} \\ (\lambda'\mu') \\ w', I' \end{matrix} \middle\| g \begin{pmatrix} \frac{1}{2} \\ (10) \\ \frac{1}{2} \end{pmatrix} \middle\| \begin{matrix} \frac{\lambda}{2} \\ (\lambda\mu) \\ w, I \end{matrix} \right\rangle.$$

This transforms into

$$\langle\langle(\lambda\mu)\|\partial\|(\lambda'\mu')\rangle\rangle\langle(\lambda'\mu')Y'(w')I';\ (01) - \frac{1}{3}\frac{1}{2}\|(\lambda\mu)Y(w)I\rangle$$

$$= (-1)^{\frac{\lambda'}{2} - \frac{\lambda}{2} + I - I'} \sqrt{\frac{(2I'+1)(\lambda'+1)}{(2I+1)(\lambda+1)}} \langle\langle(\lambda'\mu')\|g\|(\lambda\mu)\rangle\rangle$$

$$\times \langle(\lambda\mu)Y(w)I;\ (10)\frac{1}{3}\frac{1}{2}\|(\lambda'\mu')Y'(w')I'\rangle.$$

Using in addition the symmetry property under $1 \leftrightarrow 3$ interchange of both the full $SU(3)$ Wigner coefficient and the $SU(2)$ I-spin Wigner coefficient, we can relate the two $SU(3)$ reduced Wigner coefficients

$$\langle(\lambda\mu)Y(w)I;\ (10)\frac{1}{3}\frac{1}{2}\|(\lambda'\mu')Y'(w')I'\rangle$$

$$= (-1)^{\phi((\lambda\mu))+\phi((10))-\phi((\lambda'\mu'))+I'-I-\frac{1}{2}} \sqrt{\frac{dim(\lambda'\mu')(2I+1)}{dim(\lambda\mu)(2I'+1)}}$$

$$\langle(\lambda'\mu')Y'(w')I';\ (01) - \frac{1}{3}\frac{1}{2}\|(\lambda\mu)Y(w)I\rangle,$$

where we have again used $\omega((10)w = 0, \frac{1}{2}m_\ell) = \frac{1}{2} - m_\ell$. This leads to the relationship between the $SU(2) \star SU(3)$ doubly reduced matrix elements

$$\langle\langle(\lambda'\mu')\|g\|(\lambda\mu)\rangle\rangle$$

$$= (-1)^{\phi((\lambda\mu))+\phi((10))-\phi((\lambda'\mu'))+\frac{\lambda'}{2}-\frac{\lambda}{2}-\frac{1}{2}}\sqrt{\frac{(\lambda+1)dim(\lambda\mu)}{(\lambda'+1)dim(\lambda'\mu')}}\langle\langle(\lambda\mu)\|\partial\|(\lambda'\mu')\rangle\rangle.$$

With these results we are now ready to undertake the calculation of $SU(3)$ Wigner coefficients using the vector coherent state construction.

5.3. Calculation.

It will now be important to rewrite the vector coherent state construction of the $SU(3) \supset SU(2) \times U(1)$ orthonormal basis states in terms of the double (upper and lower index) notation

$$|(\lambda\mu)Y(w)IM_I\rangle = \left(K(\lambda\mu)\right)^{-1}_{w,I}\left[Z^0_{\frac{w}{2}}(\mathbf{A}) \times \left|\begin{matrix}\frac{\lambda}{2}\\(\lambda\mu)\\0,\frac{\lambda}{2}\end{matrix}\right)\right]^{\frac{\lambda}{2}\frac{\lambda}{2}}_{IM_I},$$

where the square bracket now indicates vector coupling in both upper and lower angular momenta although the upper coupling is a trivial one, since $A_j = \sum_a g^a_3\partial/\partial g^a_j$ is a $U(2)$-scalar, (with upper angular momentum 0). Also recall that we use a right to left coupling order convention: $[\frac{\lambda}{2} \times \frac{w}{2}] \rightarrow IM_I$.

To evaluate the simple $SU(3)$ Wigner coefficients for the coupling with a fundamental tensor or its conjugate, we can now consider the explicit action of such tensor operators on this state $|(\lambda\mu)Y(w)IM_I\rangle$; where $SU(2)$ coupling in the upper indices will be used to enforce a specific shift property. The calculations will be simplified by the following strategy: We can choose components of g or $\partial/\partial\mathbf{g}$ which commute with A_j, $j = 1,2$, such as

g^a_3 (with lower pattern I-spin of 0),   or

$\partial/\partial g^a_k$, $k = 1, 2$, (with lower pattern I-spin of $\frac{1}{2}$).

Case1. The Coupling $[(\lambda\mu) \times (10)] \to (\lambda'\mu') = (\lambda+1,\mu), (\lambda-1,\mu+1)$.

It turns out that the simplest case involves action with the operators $\partial/\partial g_k^a$, $(k = 1,2))$, which not only commute with $Z(\mathbf{A})$ but give very simple results when acting on extremal $U(2){\star}SU(3)$ states. It will therefore be convenient to consider first the action of shift operators built from these $\partial/\partial \mathbf{g}$ acting on $(\lambda'\mu')$ states, and then use the $1 \leftrightarrow 3$ interchange symmetry to obtain the Wigner coefficients for the coupling $[(\lambda\mu) \times (10)] \to (\lambda'\mu')$. The needed shift operations are given by

$$
\left\langle \begin{array}{c} \frac{\lambda}{2}\frac{\lambda}{2} \\ (\lambda\mu) \\ w, IM_I \end{array} \left| \left[\frac{\partial}{\partial g} \begin{pmatrix} \frac{1}{2} \\ (01) \\ \frac{1}{2} \end{pmatrix} \times \left| \begin{array}{c} \frac{\lambda'}{2} \\ (\lambda'\mu') \\ w, I' \end{array} \right\rangle \right]^{\frac{\lambda}{2}\frac{\lambda}{2}} \right._{IM_I}
$$

$$
= \langle\langle(\lambda\mu)\|\partial\|(\lambda'\mu')\rangle\rangle \; \langle(\lambda'\mu')Y'(w)I'; \; (01) - \frac{1}{3}\frac{1}{2}\|(\lambda\mu)Y(w)I\rangle,
$$

where $\frac{\lambda'}{2} = \frac{1}{2}(\lambda \pm 1)$, only. Using the vector coherent state construction for the state $(\lambda'\mu')$, the fact that the components of $\partial^{\frac{1}{2}}(01)$ commute with $Z(\mathbf{A})$, and a standard angular momentum recoupling transformation, this scalar product can be expressed by

$$
\left\langle \begin{array}{c} \frac{\lambda}{2}\frac{\lambda}{2} \\ (\lambda\mu) \\ w, IM_I \end{array} \left| \left[\frac{\partial}{\partial g} \begin{pmatrix} \frac{1}{2} \\ (01) \\ \frac{1}{2} \end{pmatrix} \times \left| \begin{array}{c} \frac{\lambda'}{2} \\ (\lambda'\mu') \\ w, I' \end{array} \right\rangle \right]^{\frac{\lambda}{2}\frac{\lambda}{2}} \right._{IM_I}
$$

$$
= \frac{1}{(K(\lambda'\mu'))_{wI'}} \left\langle \begin{array}{c} \frac{\lambda}{2}\frac{\lambda}{2} \\ (\lambda\mu) \\ w, IM_I \end{array} \left| \left[Z_{\frac{w}{2}}^0(\mathbf{A}) \times \left[\frac{\partial}{\partial g} \begin{pmatrix} \frac{1}{2} \\ (01) \\ \frac{1}{2} \end{pmatrix} \times \left| \begin{array}{c} \frac{\lambda'}{2} \\ (\lambda'\mu') \\ 0, \frac{\lambda'}{2} \end{array} \right\rangle \right]^{\frac{\lambda'}{2}} \right]^{\frac{\lambda}{2}\frac{\lambda}{2}} \right._{IM_I}
$$

$$
\times (-1)^{\frac{\lambda'}{2}+\frac{w}{2}-I'} U\left(\frac{w}{2}\frac{\lambda'}{2}I\frac{1}{2}; I'\frac{\lambda}{2}\right)(-1)^{I-\frac{\lambda}{2}-\frac{w}{2}}.
$$

The first phase factor comes from the change in the coupling order $[\frac{\lambda'}{2} \times \frac{w}{2}]I' \to [\frac{w}{2} \times \frac{\lambda'}{2}]I'$, the second from the reordering of $[\frac{w}{2} \times \frac{\lambda}{2}]I \to [\frac{\lambda}{2} \times \frac{w}{2}]I$ subsequent to the recoupling transformation. Note also that no sum is needed in the recoupling transformation: The action of $\partial_{\frac{1}{2}}^{\frac{1}{2}}$ on an extremal state of $(\lambda'\mu')$ with $w=0$ and $Y' = \frac{1}{3}(\lambda' + 2\mu')$ creates a state with $Y = \frac{1}{3}(\lambda' + 2\mu' - 1) = \frac{1}{3}(\lambda + 2\mu)$, again an extremal state with $w=0$, so that the lower pattern angular momentum $\frac{\lambda}{2}$ is fixed uniquely. Note also that the upper pattern recoupling is unique because of the upper angular momentum of 0 for $Z(\mathbf{A})$. The uniqueness of the action of $\partial_{\frac{1}{2}}^{\frac{1}{2}}$ on the extremal state of $(\lambda'\mu')$ leads to

$$
\left[\frac{\partial}{\partial g} \begin{pmatrix} \frac{1}{2} \\ (01) \\ \frac{1}{2} \end{pmatrix} \left| \begin{matrix} \frac{\lambda'}{2} \\ (\lambda'\mu') \\ 0, \frac{\lambda'}{2} \end{matrix} \right\rangle \right]_{\frac{\lambda}{2}m}^{\frac{\lambda}{2}\frac{\lambda}{2}}
$$

$$
= \left| \begin{matrix} \frac{\lambda}{2}\frac{\lambda}{2} \\ (\lambda\mu) \\ 0, \frac{\lambda}{2}m \end{matrix} \right\rangle \left\langle \begin{matrix} \frac{\lambda}{2} \\ (\lambda\mu) \\ 0, \frac{\lambda}{2} \end{matrix} \right\| \frac{\partial}{\partial g} \begin{pmatrix} \frac{1}{2} \\ (01) \\ \frac{1}{2} \end{pmatrix} \left\| \begin{matrix} \frac{\lambda'}{2} \\ (\lambda'\mu') \\ 0, \frac{\lambda'}{2} \end{matrix} \right\rangle,
$$

where

$$
\left\langle \begin{matrix} \frac{\lambda}{2} \\ (\lambda\mu) \\ 0, \frac{\lambda}{2} \end{matrix} \right\| \frac{\partial}{\partial g} \begin{pmatrix} \frac{1}{2} \\ (01) \\ \frac{1}{2} \end{pmatrix} \left\| \begin{matrix} \frac{\lambda'}{2} \\ (\lambda'\mu') \\ 0, \frac{\lambda'}{2} \end{matrix} \right\rangle
$$

$$
= \left\langle \begin{matrix} \frac{\lambda'}{2} \\ (\lambda'\mu') \\ 0, \frac{\lambda'}{2} \end{matrix} \right\| g \begin{pmatrix} \frac{1}{2} \\ (10) \\ \frac{1}{2} \end{pmatrix} \left\| \begin{matrix} \frac{\lambda}{2} \\ (\lambda\mu) \\ \frac{\lambda}{2} \end{matrix} \right\rangle (-1)^{\lambda'-\lambda-1} \frac{\lambda'+1}{\lambda+1}
$$

$$
= \frac{\lambda'+1}{\lambda+1} \langle\langle(\lambda'\mu')\|\mathbf{g}\|(\lambda\mu)\rangle\rangle \; \langle(\lambda\mu)\frac{1}{3}(\lambda+2\mu)\frac{\lambda}{2}; \, (10)\frac{1}{3}\frac{1}{2}\|(\lambda'\mu')\frac{1}{3}(\lambda+2\mu+1)\frac{\lambda'}{2}\rangle
$$

$$
= \frac{\lambda'+1}{\lambda+1} \langle\langle(\lambda'\mu')\|\mathbf{g}\|(\lambda\mu)\rangle\rangle \times 1.
$$

We have used the relationship between the double-barred matrix elements of $\partial/\partial\mathbf{g}$ and $\mathbf{g}$ in the first step and the fact that the $SU(3)$ reduced Wigner coefficient for the unique coupling between extremal states has the value $+1$, (generalized Condon-Shortley phase convention). Note also that $\lambda' - \lambda - 1 =$ even for both possible values of λ'.

By substituting this result for the action of $\partial_{\frac{1}{2}}^{\frac{1}{2}}$ on the extremal state into the state vector construction, and by converting

$$\left[Z_{\frac{w}{2}}^0(\mathbf{A}) \times \left| \begin{matrix} \frac{\lambda}{2} \\ (\lambda\mu) \\ 0, \frac{\lambda}{2} \end{matrix} \right\rangle \right]_{IM_I}^{\frac{\lambda}{2}\frac{\lambda}{2}} = \left(K(\lambda\mu) \right)_{wI} \left| \begin{matrix} \frac{\lambda}{2}\frac{\lambda}{2} \\ (\lambda\mu) \\ w, IM_I \end{matrix} \right\rangle$$

into a normalized state with the K-factor, we obtain

$$\langle\langle(\lambda\mu)\|\partial\|(\lambda'\mu')\rangle\rangle\, \langle(\lambda'\mu')Y'(w)I';\, (01) - \frac{1}{3}\frac{1}{2}\|(\lambda\mu)Y(w)I\rangle$$

$$= \frac{\lambda'+1}{\lambda+1}\langle\langle(\lambda'\mu')\|\mathbf{g}\|(\lambda\mu)\rangle\rangle \frac{\left(K(\lambda\mu)\right)_{wI}}{\left(K(\lambda'\mu')\right)_{wI'}}(-1)^{\frac{\lambda'}{2}-\frac{\lambda}{2}+I-I'}U\left(\frac{w}{2}\frac{\lambda'}{2}I\frac{1}{2};I'\frac{\lambda}{2}\right).$$

Using the relation between $\langle\langle(\lambda\mu)\|\partial\|(\lambda'\mu')\rangle\rangle$ and $\langle\langle(\lambda'\mu')\|\mathbf{g}\|(\lambda\mu)\rangle\rangle$, this gives

$$\langle(\lambda'\mu')Y'(w)I';\, (01) - \frac{1}{3}\frac{1}{2}\|(\lambda\mu)Y(w)I\rangle$$

$$= \sqrt{\frac{(\lambda'+1)dim(\lambda\mu)}{(\lambda+1)dim(\lambda'\mu')}}\left[\frac{\left(K(\lambda\mu)\right)_{wI}}{\left(K(\lambda'\mu')\right)_{wI'}}\right](-1)^{I+\frac{1}{2}-I'+\phi(\lambda'\mu')-\phi(\lambda\mu)-\phi(01)}$$

$$\times U\left(\frac{w}{2}\frac{\lambda'}{2}I\frac{1}{2};I'\frac{\lambda}{2}\right).$$

Finally the symmetry property under $1 \leftrightarrow 3$ interchange for the $SU(3)$ reduced Wigner coefficient, (full $SU(3)$ Wigner coefficient divided by I-spin $SU(2)$ Wigner coefficient), gives

Result I. For $(\lambda'\mu') = (\lambda+1,\mu), (\lambda-1,\mu+1)$:

$$\langle(\lambda\mu)Y(w)I;\ (10)\tfrac{1}{3}\tfrac{1}{2}\|(\lambda'\mu')Y'(w)I'\rangle$$

$$= \sqrt{\frac{(\lambda'+1)(2I+1)}{(\lambda+1)(2I'+1)}}\left[\frac{(K(\lambda\mu))_{wI}}{(K(\lambda'\mu'))_{wI'}}\right]U\left(\frac{w}{2}\frac{\lambda'}{2}I\frac{1}{2};I'\frac{\lambda}{2}\right).$$

We see therefore that this $SU(3)$ reduced Wigner coefficient is given by simple K and dimensional factors and an $SU(2)$ recoupling coefficient. The $SU(3)$ Wigner coefficient is thus known via a simple $SU(2)$ coefficient.

In exactly the same manner, using a shift operator built from g_3^a, (with lower pattern I-spin of 0),

$$\left\langle\begin{array}{c}\tfrac{\lambda'}{2}\tfrac{\lambda'}{2}\\(\lambda'\mu')\\w,IMI\end{array}\right\|\left[\begin{array}{c}g\left(\begin{array}{c}\tfrac{1}{2}\\(10)\\0\end{array}\right)\times\left|\begin{array}{c}\tfrac{\lambda}{2}\\(\lambda\mu)\\w,I\end{array}\right\rangle\end{array}\right]^{\tfrac{\lambda'}{2}\tfrac{\lambda'}{2}}_{IMI}$$

$$= \langle\langle(\lambda'\mu')\|\mathbf{g}\|(\lambda\mu)\rangle\rangle\ \langle(\lambda\mu)Y(w)I;\ (10)-\tfrac{2}{3}0\|(\lambda'\mu')Y'(w)I\rangle$$

$$= \frac{1}{(K(\lambda\mu))_{wI}}\left\langle\begin{array}{c}\tfrac{\lambda'}{2}\tfrac{\lambda'}{2}\\(\lambda'\mu')\\w,IMI\end{array}\right\|\left[Z^0_{\tfrac{w}{2}}(\mathbf{A})\times\left[g\left(\begin{array}{c}\tfrac{1}{2}\\(10)\\0\end{array}\right)\times\left|\begin{array}{c}\tfrac{\lambda}{2}\\(\lambda\mu)\\0,\tfrac{\lambda}{2}\end{array}\right\rangle\right]^{\tfrac{\lambda'}{2}}\right]^{\tfrac{\lambda'}{2}\tfrac{\lambda'}{2}}_{IMI},$$

where we have again used the fact that g_3^a commutes with $Z(\mathbf{A})$, and where *both* lower and upper recouplings are now unique because there is one 0 angular momentum in *both* lower and upper coupling schemes. The action of $g_0^{\frac{1}{2}}(10)$ on the extremal state of $(\lambda\mu)$, with $w=0$, is now a little more complicated. This action makes a state with $Y=\frac{1}{3}(\lambda+2\mu)-\frac{2}{3}=\frac{1}{3}(\lambda'+2\mu')-1$ for both

$(\lambda'\mu') = (\lambda+1,\mu), (\lambda-1,\mu+1)$; ie a state with $w' = 1$.

$$\left[g\begin{pmatrix} \frac{1}{2} \\ (10) \\ 0 \end{pmatrix} \times \left| \begin{matrix} \frac{\lambda}{2} \\ (\lambda\mu) \\ 0, \frac{\lambda}{2} \end{matrix} \right\rangle \right]^{\frac{\lambda'}{2}\frac{\lambda'}{2}}_{\frac{\lambda}{2}m}$$

$$= \left| \begin{matrix} \frac{\lambda'}{2}\frac{\lambda'}{2} \\ (\lambda'\mu') \\ w'=1, I=\frac{\lambda}{2}, m \end{matrix} \right\rangle \left\langle \begin{matrix} \frac{\lambda'}{2}\frac{\lambda'}{2} \\ (\lambda'\mu') \\ 1, \frac{\lambda}{2}m \end{matrix} \right| \left[g\begin{pmatrix} \frac{1}{2} \\ (10) \\ 0 \end{pmatrix} \times \left| \begin{matrix} \frac{\lambda}{2} \\ (\lambda\mu) \\ 0, \frac{\lambda}{2} \end{matrix} \right\rangle \right]^{\frac{\lambda'}{2}\frac{\lambda'}{2}}_{\frac{\lambda}{2}m}.$$

To evaluate the m-independent overlap, we   ·

(1) take the hermitian conjugate of the explicit vector coherent state construction for the $w = 1$ state

$$\left\langle \begin{matrix} \frac{\lambda'}{2}\frac{\lambda'}{2} \\ (\lambda'\mu') \\ 1, \frac{\lambda}{2}m \end{matrix} \right| =$$

$$\frac{1}{(K(\lambda'\mu'))_{1\frac{\lambda}{2}}} \sum_{i,m(i)} (-1)^{\frac{1}{2}-m(i)} \langle \frac{\lambda'}{2}(m-m(i))\, \frac{1}{2}-m(i)|\frac{\lambda}{2}m\rangle \left\langle \begin{matrix} \frac{\lambda'}{2}\frac{\lambda'}{2} \\ (\lambda'\mu') \\ 0, \frac{\lambda'}{2}(m-m(i)) \end{matrix} \right| B_i$$

with $m(i) = +\frac{1}{2}, -\frac{1}{2}$, for $i = 1, 2$;

(2) use the commutator

$$[B_i, g_3^a] = g_i^a$$

to bring the operator B_i to the right;

(3) use the fact that B_i annihilates the extremal state $(\lambda\mu)$, with $w = 0$; and

(4) use the symmetry property under $1 \leftrightarrow 3$ interchange for the $SU(2)$ Wigner coefficient, as well as the m'-independence of the final overlap matrix element.

This then gives

$$
\left\langle
\begin{array}{c}
\frac{\lambda'}{2}\frac{\lambda'}{2} \\
(\lambda'\mu') \\
1,\frac{\lambda}{2}m
\end{array}
\middle|
\left[
g
\left(
\begin{array}{c}
\frac{1}{2} \\
(10) \\
0
\end{array}
\right)
\times
\left|
\begin{array}{c}
\frac{\lambda}{2}\frac{\lambda}{2} \\
(\lambda\mu) \\
0,\frac{\lambda}{2}
\end{array}
\right\rangle
\right]^{\frac{\lambda'}{2}\frac{\lambda'}{2}}_{\frac{\lambda}{2}m}
\right.
$$

$$
= \frac{1}{(K(\lambda'\mu'))_{1,\frac{\lambda}{2}}}(-1)^{\frac{\lambda'}{2}-\frac{1}{2}-\frac{\lambda}{2}}\sqrt{\frac{\lambda'+1}{\lambda+1}}
\left\langle
\begin{array}{c}
\frac{\lambda'}{2}\frac{\lambda'}{2} \\
(\lambda'\mu') \\
0,\frac{\lambda'}{2}m'
\end{array}
\middle|
\left[
g
\left(
\begin{array}{c}
\frac{1}{2} \\
(10) \\
\frac{1}{2}
\end{array}
\right)
\times
\left|
\begin{array}{c}
\frac{\lambda}{2} \\
(\lambda\mu) \\
0,\frac{\lambda}{2}
\end{array}
\right\rangle
\right]^{\frac{\lambda'}{2}\frac{\lambda'}{2}}_{\frac{\lambda'}{2}m'}
\right.
$$

$$
= \frac{1}{(K(\lambda'\mu'))_{1,\frac{\lambda}{2}}}(-1)^{\frac{\lambda'}{2}-\frac{1}{2}-\frac{\lambda}{2}}\sqrt{\frac{\lambda'+1}{\lambda+1}}\langle\langle(\lambda'\mu')\|\mathbf{g}\|(\lambda\mu)\rangle\rangle \times 1.
$$

In the last step we have used the trivial value for the reduced Wigner coefficient, $\langle(\lambda\mu)\frac{1}{3}(\lambda+2\mu)\frac{\lambda}{2};\ (10)\frac{1}{3}\frac{1}{2}\|(\lambda'\mu')\frac{1}{3}(\lambda'+2\mu')\frac{\lambda'}{2}\rangle = +1$, for this unique coupling between extremal states. With this result we get

$$
\langle\langle(\lambda'\mu')\|\mathbf{g}\|(\lambda\mu)\rangle\rangle\ \langle(\lambda\mu)Y(w)I;\ (10)-\frac{2}{3}0\|(\lambda'\mu')Y'(w)I\rangle
$$

$$
= \frac{1}{(K(\lambda\mu))_{wI}}
\left\langle
\begin{array}{c}
\frac{\lambda'}{2}\frac{\lambda'}{2} \\
(\lambda'\mu') \\
w,IM_I
\end{array}
\middle|
\left[
Z^0_{\frac{w}{2}}(\mathbf{A})
\times
\left|
\begin{array}{c}
\frac{\lambda'}{2}\frac{\lambda'}{2} \\
(\lambda'\mu') \\
1,\frac{\lambda}{2}
\end{array}
\right\rangle
\right]^{\frac{\lambda'}{2}\frac{\lambda'}{2}}_{IM_I}
\right.
$$

$$
\times(-1)^{\frac{\lambda'}{2}+\frac{1}{2}-\frac{\lambda}{2}}\frac{1}{(K(\lambda'\mu'))_{1\frac{\lambda}{2}}}\sqrt{\frac{\lambda'+1}{\lambda+1}}\langle\langle(\lambda'\mu')\|\mathbf{g}\|(\lambda\mu)\rangle\rangle.
$$

Now we can again use the explicit vector coherent state construction to express

$$
\left[
Z^0_{\frac{w}{2}}(\mathbf{A})
\times
\left|
\begin{array}{c}
\frac{\lambda'}{2} \\
(\lambda'\mu') \\
1,\frac{\lambda}{2}
\end{array}
\right\rangle
\right]^{\frac{\lambda'}{2}\frac{\lambda'}{2}}_{IM_I}
$$

$$= \frac{1}{(K(\lambda'\mu'))_{1\frac{\lambda}{2}}} \left[Z^0_{\frac{w}{2}}(\mathbf{A}) \times \left[\mathbf{A}^0_{\frac{1}{2}} \times \left| (\lambda'\mu') \begin{array}{c} \frac{\lambda'}{2} \\ \\ 0, \frac{\lambda'}{2} \end{array} \right\rangle \right]^{\frac{\lambda'}{2}}_{\frac{\lambda'}{2}} \right]^{\frac{\lambda'}{2}\frac{\lambda'}{2}}_{IM_I}$$

$$= \frac{\sqrt{(w+1)}}{(K(\lambda'\mu'))_{1\frac{\lambda}{2}}} U\left(\frac{\lambda'}{2} \frac{1}{2} I \frac{w}{2}; \frac{\lambda}{2} \frac{w+1}{2} \right) \left[Z^0_{\frac{w+1}{2}}(\mathbf{A}) \times \left| (\lambda'\mu') \begin{array}{c} \frac{\lambda'}{2} \\ \\ 0, \frac{\lambda'}{2} \end{array} \right\rangle \right]^{\frac{\lambda'}{2}\frac{\lambda'}{2}}_{IM_I},$$

where we have used the basic property of the symmetric polynomials, $Z(\mathbf{A})$

$$\left[Z^0_{\frac{w}{2}}(\mathbf{A}) \times \mathbf{A}^0_{\frac{1}{2}} \right]^0_{\frac{w'}{2}} = \sqrt{w+1} \delta_{\frac{w'}{2}, \frac{w+1}{2}} Z^0_{\frac{w+1}{2}}(\mathbf{A}).$$

Substituting this result into the above yields

Result II.

$$\langle (\lambda\mu) Y(w) I; \ (10) - \frac{2}{3} 0 \| (\lambda'\mu') Y'(w+1) I \rangle$$

$$= \left[\frac{(K(\lambda'\mu'))_{w+1,I}}{(K(\lambda\mu))_{wI}} \right] (-1)^{\frac{\lambda'}{2} - \frac{1}{2} - \frac{\lambda}{2}} \sqrt{\frac{\lambda'+1}{\lambda+1}} \left[\frac{\sqrt{w+1}}{(K(\lambda'\mu'))^2_{1\frac{\lambda}{2}}} \right] U\left(\frac{\lambda'}{2} \frac{1}{2} I \frac{w}{2}; \frac{\lambda}{2} \frac{w+1}{2} \right).$$

Again, the $SU(3)$ reduced Wigner coefficient is given by simple K and dimensional factors and a very simple $SU(2)$ Racah coefficient. Note that results I and II apply to the coupling $[(\lambda\mu) \times (10)](\lambda'\mu')$ with $(\lambda'\mu') = (\lambda+1, \mu), (\lambda-1, \mu+1)$ only.

Case 2. The Coupling $[(\lambda\mu) \times (10)] \rightarrow (\lambda, \mu-1)$.

To evaluate the Wigner coefficients with $(\lambda'\mu') = (\lambda, \mu-1)$, obtainable via the removal of one square each from both rows 1 and 2 of the two-rowed tableau $[\lambda + \mu, \mu]$, we consider first the shift tensor

$$\left[\frac{\partial}{\partial g} \begin{pmatrix} \frac{1}{2} \\ (01) \\ \frac{1}{2} \end{pmatrix} \times \frac{\partial}{\partial g} \begin{pmatrix} \frac{1}{2} \\ (01) \\ \frac{1}{2} \end{pmatrix} \right]^{00}_{00},$$

with both lower and upper pattern angular momenta coupled to zero. Note that this is an $SU(3)$ (10)-tensor, with $Y = -\frac{2}{3}$. Thus

$$
\left\langle \begin{matrix} \frac{\lambda}{2}\frac{\lambda}{2} \\ (\lambda, \mu - 1) \\ w, IM_I \end{matrix} \middle\| \left[\left[\frac{\partial}{\partial g} \begin{pmatrix} \frac{1}{2} \\ (01) \\ \frac{1}{2} \end{pmatrix} \times \frac{\partial}{\partial g} \begin{pmatrix} \frac{1}{2} \\ (01) \\ \frac{1}{2} \end{pmatrix} \right]_0^0 \times \middle| \begin{matrix} \frac{\lambda}{2} \\ (\lambda\mu) \\ w, I \end{matrix} \right\rangle \right]_{IM_I}^{\frac{\lambda}{2}\frac{\lambda}{2}}
$$

$$
= \langle (\lambda\mu)Y(w)I;\ (10) - \tfrac{2}{3}0 \| (\lambda, \mu - 1)Y'(w)I \rangle \langle\langle (\lambda, \mu - 1) \| [\partial \times \partial]^{(10)} \| (\lambda\mu) \rangle\rangle,
$$

using an obvious shorthand notation for the operator in the $U(2) \star SU(3)$ doubly reduced matrix element. Now using the vector coherent state construction of the ket and the fact that $\partial/\partial g$ operators with lower pattern I-spins of $\frac{1}{2}$ commute with $Z(\mathbf{A})$, the above overlap integral becomes

$$
\left\langle \begin{matrix} \frac{\lambda}{2}\frac{\lambda}{2} \\ (\lambda, \mu - 1) \\ w, IM_I \end{matrix} \middle\| \left[Z_{\frac{w}{2}}^0(\mathbf{A}) \times \left[\left[\frac{\partial}{\partial g} \begin{pmatrix} \frac{1}{2} \\ (01) \\ \frac{1}{2} \end{pmatrix} \times \frac{\partial}{\partial g} \begin{pmatrix} \frac{1}{2} \\ (01) \\ \frac{1}{2} \end{pmatrix} \right]_0^0 \times \middle| \begin{matrix} \frac{\lambda}{2} \\ (\lambda\mu) \\ 0, \frac{\lambda}{2} \end{matrix} \right\rangle \right]^{\frac{\lambda}{2}} \right]_{IM_I}^{\frac{\lambda}{2}\frac{\lambda}{2}}
$$

$$
\times [(K(\lambda\mu))_{wI}]^{-1}
$$

$$
= \frac{1}{(K(\lambda\mu))_{wI}} \left\langle \begin{matrix} \frac{\lambda}{2}\frac{\lambda}{2} \\ (\lambda, \mu - 1) \\ w, IM_I \end{matrix} \middle\| \left[Z_{\frac{w}{2}}^0(\mathbf{A}) \times \middle| \begin{matrix} \frac{\lambda}{2} \\ (\lambda, \mu - 1) \\ 0, \frac{\lambda}{2} \end{matrix} \right\rangle \right]_{IM_I}^{\frac{\lambda}{2}\frac{\lambda}{2}} \times
$$

$$
\langle (\lambda\mu)\tfrac{1}{3}(\lambda+2\mu)\tfrac{\lambda}{2};\ (10) - \tfrac{2}{3}0 \| (\lambda, \mu-1)\tfrac{1}{3}(\lambda+2\mu-2)\tfrac{\lambda}{2} \rangle \langle\langle (\lambda, \mu-1) \| [\partial \times \partial]^{(10)} \| (\lambda\mu) \rangle\rangle.
$$

The $SU(3)$ reduced Wigner coefficient in the last line is related to a trivial one with value $+1$ by a $1 \leftrightarrow 3$ interchange symmetry

$$
\langle (\lambda\mu)\tfrac{1}{3}(\lambda + 2\mu)\tfrac{\lambda}{2};\ (10) - \tfrac{2}{3}0 \| (\lambda, \mu - 1)\tfrac{1}{3}(\lambda + 2\mu - 2)\tfrac{\lambda}{2} \rangle
$$

$$= \sqrt{\frac{dim(\lambda, \mu - 1)}{dim(\lambda\mu)}} \langle (\lambda, \mu - 1)\tfrac{1}{3}(\lambda + 2\mu - 2)\tfrac{\lambda}{2}; \ (01)\tfrac{2}{3}0\|(\lambda\mu)\tfrac{1}{3}(\lambda + 2\mu)\tfrac{\lambda}{2}\rangle$$

$$= \sqrt{\frac{dim(\lambda, \mu - 1)}{dim(\lambda\mu)}} \times 1.$$

With this result the above vector coherent state expansion at once leads to the desired

Result III.

$$\langle (\lambda\mu)Y(w)I; \ (10) - \tfrac{2}{3}0\|(\lambda, \mu - 1)Y'(w)\tfrac{\lambda}{2}\rangle$$

$$= \left[\frac{(K(\lambda, \mu - 1))_{wI}}{(K(\lambda\mu))_{wI}} \right] \sqrt{\frac{\mu(\lambda + \mu + 1)}{(\mu + 1)(\lambda + \mu + 2)}}.$$

For the final $SU(3)$ reduced Wigner coefficient we use the fact that the operator

$$\left[g \begin{pmatrix} \tfrac{1}{2} \\ (10) \\ 0 \end{pmatrix} \times g \begin{pmatrix} \tfrac{1}{2} \\ (10) \\ \tfrac{1}{2} \end{pmatrix} \right]^{0}_{\frac{1}{2}m(i)} = \frac{1}{\sqrt{2}}(g_3^1 g_i^2 - g_i^1 g_3^2)$$

(with $i = 1, 2$ for $m(i) = +\tfrac{1}{2}, -\tfrac{1}{2}$), is an $SU(3)$ (01)-tensor with the properties:

(1) It is a shift operator which adds one square each to rows 1 and 2 of a Young tableau.

(2) It commutes with $A_i = \sum_a g_3^a \partial/\partial g_i^a$.

Then, using the shift operator property and the vector coherent state expansion of the ket

$$\left\langle \begin{matrix} \tfrac{\lambda}{2}\tfrac{\lambda}{2} \\ (\lambda\mu) \\ w, IM_I \end{matrix} \right| \left[\left[g \begin{pmatrix} \tfrac{1}{2} \\ (10) \\ 0 \end{pmatrix} \times g \begin{pmatrix} \tfrac{1}{2} \\ (10) \\ \tfrac{1}{2} \end{pmatrix} \right]^{0}_{\frac{1}{2}} \times \left| \begin{matrix} \tfrac{\lambda}{2} \\ (\lambda, \mu - 1) \\ w', I' \end{matrix} \right\rangle \right]^{\tfrac{\lambda}{2}\tfrac{\lambda}{2}}_{IM_I}$$

$$= \langle (\lambda, \mu-1)Y'(w')I'; (01) -\tfrac{1}{3}\tfrac{1}{2} \| (\lambda\mu)Y(w)I \rangle \langle\langle (\lambda\mu) \| [g \times g]^{(01)} \| (\lambda, \mu-1) \rangle\rangle$$

$$= \frac{1}{(K(\lambda, \mu-1))_{w'I'}} \sum_{\frac{\lambda''}{2}} (-1)^{\frac{\lambda}{2}+\frac{w'}{2}-I'} U\left(\frac{w'}{2}\frac{\lambda}{2}I'\frac{1}{2}; I'\frac{\lambda''}{2}\right)(-1)^{I+\frac{\lambda''}{2}-\frac{w'}{2}}$$

$$\times \left\langle \begin{matrix} \tfrac{\lambda}{2}\tfrac{\lambda}{2} \\ (\lambda\mu) \\ w, IM_I \end{matrix} \;\middle\|\; \left[Z^0_{\frac{w'}{2}}(\mathbf{A}) \times \left[\left[g\begin{pmatrix} \tfrac{1}{2} \\ (10) \\ 0 \end{pmatrix} \times g\begin{pmatrix} \tfrac{1}{2} \\ (10) \\ \tfrac{1}{2} \end{pmatrix} \right]^0_{\frac{1}{2}} \times \middle| \begin{matrix} \tfrac{\lambda}{2} \\ (\lambda, \mu-1) \\ 0, \tfrac{\lambda}{2} \end{matrix} \right\rangle \right]^{\frac{\lambda}{2}} \right]^{\frac{\lambda}{2}\frac{\lambda}{2}}_{\frac{\lambda''}{2}} \right\rangle_{IM_I} .$$

We now use the fact that

$$\left[\left[g\begin{pmatrix} \tfrac{1}{2} \\ (10) \\ 0 \end{pmatrix} \times g\begin{pmatrix} \tfrac{1}{2} \\ (10) \\ \tfrac{1}{2} \end{pmatrix} \right]^0_{\frac{1}{2}} \middle| \begin{matrix} \tfrac{\lambda}{2} \\ (\lambda, \mu-1) \\ 0, \tfrac{\lambda}{2} \end{matrix} \right\rangle \right]^{\frac{\lambda}{2}\frac{\lambda}{2}}_{\frac{\lambda''}{2}m''} = \left| \begin{matrix} \tfrac{\lambda}{2}\tfrac{\lambda}{2} \\ (\lambda\mu) \\ w=1, \tfrac{\lambda''}{2}m'' \end{matrix} \right\rangle \alpha_{\lambda''},$$

where $\alpha_{\lambda''}$ can be calculated from the overlap

$$\alpha_{\lambda''} = \left\langle \begin{matrix} \tfrac{\lambda}{2}\tfrac{\lambda}{2} \\ (\lambda\mu) \\ w=1, \tfrac{\lambda''}{2}m'' \end{matrix} \;\middle\|\; \left[\left[g\begin{pmatrix} \tfrac{1}{2} \\ (10) \\ 0 \end{pmatrix} \times g\begin{pmatrix} \tfrac{1}{2} \\ (10) \\ \tfrac{1}{2} \end{pmatrix} \right]^0_{\frac{1}{2}} \times \middle| \begin{matrix} \tfrac{\lambda}{2} \\ (\lambda, \mu-1) \\ 0, \tfrac{\lambda}{2} \end{matrix} \right\rangle \right]^{\frac{\lambda}{2}\frac{\lambda}{2}}_{\frac{\lambda''}{2}m''}$$

$$= \frac{1}{(K(\lambda\mu))_{1\frac{\lambda''}{2}}} \left\langle \begin{matrix} \tfrac{\lambda}{2}\tfrac{\lambda}{2} \\ (\lambda\mu) \\ 0, \tfrac{\lambda}{2}m \end{matrix} \;\middle\|\; \left[\left[g\begin{pmatrix} \tfrac{1}{2} \\ (10) \\ \tfrac{1}{2} \end{pmatrix} \times g\begin{pmatrix} \tfrac{1}{2} \\ (10) \\ \tfrac{1}{2} \end{pmatrix} \right]^0_0 \times \middle| \begin{matrix} \tfrac{\lambda}{2} \\ (\lambda, \mu-1) \\ 0, \tfrac{\lambda}{2} \end{matrix} \right\rangle \right]^{\frac{\lambda}{2}\frac{\lambda}{2}}_{\frac{\lambda}{2}m}$$

$$= \frac{1}{(K(\lambda\mu))_{1\frac{\lambda''}{2}}} \times \langle\langle (\lambda\mu) \| [g \times g]^{(01)} \| (\lambda, \mu-1) \rangle\rangle \times 1,$$

where, in the first step, we have

(1) used the explicit vector coherent state construction of the state with $w=1$,

(2) the commutator relation

$$\left[\mathbf{B}^0_{\frac12}, \left[g\begin{pmatrix} \frac12 \\ (10) \\ 0 \end{pmatrix} \times g\begin{pmatrix} \frac12 \\ (10) \\ \frac12 \end{pmatrix} \right]^0_{\frac12} \right]^0_{JM} = \delta_{J0}\left[g\begin{pmatrix} \frac12 \\ (10) \\ \frac12 \end{pmatrix} \times g\begin{pmatrix} \frac12 \\ (10) \\ \frac12 \end{pmatrix} \right]^0_0, \quad \text{and}$$

(3) the fact that $\mathbf{B}$ annihilates the extremal state with $w = 0$. Note also that the $SU(3)$ reduced Wigner coefficient, needed in the second step, corresponds to a unique coupling between extremal states and thus has the value $+1$.

Finally, using

$$\left[Z^0_{\frac{w'}{2}}(\mathbf{A}) \times \left| \begin{matrix} \frac{\lambda}{2} \\ (\lambda\mu) \\ 1, \frac{\lambda''}{2} \end{matrix} \right\rangle \right]^{\frac{\lambda}{2}\frac{\lambda}{2}}_{IM_I} = \frac{1}{(K(\lambda\mu))_{1\frac{\lambda''}{2}}} \left[Z^0_{\frac{w'}{2}}(\mathbf{A}) \times \left[\mathbf{A}^0_{\frac12} \times \left| \begin{matrix} \frac{\lambda}{2} \\ (\lambda\mu) \\ 0, \frac{\lambda}{2} \end{matrix} \right\rangle \right]^{\frac{\lambda}{2}}_{\frac{\lambda''}{2}} \right]^{\frac{\lambda}{2}\frac{\lambda}{2}}_{IM_I}$$

$$= \frac{1}{(K(\lambda\mu))_{1\frac{\lambda''}{2}}} U\left(\frac{\lambda}{2}\frac12 I\frac{w'}{2}; \frac{\lambda''}{2}\frac{w'+1}{2}\right)\sqrt{w'+1}\left[Z^0_{\frac{w'+1}{2}}(\mathbf{A}) \times \left| \begin{matrix} \frac{\lambda}{2} \\ (\lambda\mu) \\ 0, \frac{\lambda}{2} \end{matrix} \right\rangle \right]^{\frac{\lambda}{2}\frac{\lambda}{2}}_{IM_I}$$

$$= \left[\frac{(K(\lambda\mu))_{w'+1,I}}{(K(\lambda\mu))_{1\frac{\lambda''}{2}}} \right] U\left(\frac{\lambda}{2}\frac12 I\frac{w'}{2}; \frac{\lambda''}{2}\frac{w'+1}{2}\right)\sqrt{w'+1} \left| \begin{matrix} \frac{\lambda}{2}\frac{\lambda}{2} \\ (\lambda\mu) \\ w'+1, IM_I \end{matrix} \right\rangle,$$

we obtain

$$\langle (\lambda, \mu-1)Y'(w')I'; (01) - \tfrac11\tfrac12 \| (\lambda\mu)Y(w=w'+1)I \rangle$$

$$= \left[\frac{(K(\lambda\mu))_{w'+1,I}}{(K(\lambda,\mu-1))_{w'I'}} \right] \sum_{\frac{\lambda''}{2}} \left[\frac{\sqrt{w'+1}}{(K(\lambda\mu))^2_{1\frac{\lambda''}{2}}} \right] U\left(\frac{w'}{2}\frac{\lambda}{2} I\frac12; I'\frac{\lambda''}{2}\right)$$

$$\times U\left(\frac{\lambda}{2}\frac12 I\frac{w'}{2}; \frac{\lambda''}{2}\frac{w'+1}{2}\right)(-1)^{\frac{\lambda}{2}-\frac{\lambda''}{2}+I-I'},$$

and, using the $1 \leftrightarrow 3$ interchange symmetry of the $SU(3)$ reduced Wigner coefficient, we obtain the final coefficient, (renaming $w' \to w$),

Result IV.

$$\langle(\lambda\mu)Y(w+1)I;\ (10)\frac{1}{3}\frac{1}{2}\|(\lambda,\mu-1)Y'(w)I'\rangle$$

$$=\sqrt{\frac{dim(\lambda,\mu-1)(2I+1)}{dim(\lambda\mu)(2I'+1)}}\sum_{\frac{\lambda''}{2}}(-1)^{\frac{\lambda}{2}+\frac{1}{2}-\frac{\lambda''}{2}}\left[\frac{\sqrt{w+1}}{(K(\lambda\mu))^2_{1\frac{\lambda''}{2}}}\right]\left[\frac{(K(\lambda\mu))_{w+1,I}}{(K(\lambda,\mu-1))_{wI'}}\right]$$

$$\times U\left(\frac{w}{2}\frac{\lambda}{2}I\frac{1}{2};I'\frac{\lambda''}{2}\right)U\left(\frac{\lambda}{2}\frac{1}{2}I\frac{w}{2};\frac{\lambda''}{2}\frac{w+1}{2}\right).$$

Results I - IV can now be used to construct the full table of $SU(3)$ reduced Wigner coefficients for the coupling of $(\lambda\mu)$ with the fundamental tensor (10). Only very simple $SU(2)$ Racah coefficients are needed, (all of them with one angular momentum of $\frac{1}{2}$). These have a very simple algebraic structure and lead at once to the final results as given, *eg*, in table 1 of [Vergados 1968]. (Note, however, that this tabulation uses a parameterization p,q somewhat different from the present one. We also call attention to a trivial printing error in this table: the numerator factor $(\mu-2)$ should be replaced by $(\mu-q)$ in the entry for $(\lambda'\mu')=(\lambda,\mu-1),\Lambda_1=\Lambda+\frac{1}{2}$.)

6. An Indirect Application of Vector Coherent State Theory: Construction of a Group Theoretically Sound Orthonormal Wigner Supermultiplet Basis.

In this last application it will be shown how the vector coherent state construction can be used indirectly to solve an internal labelling problem, one of the most common problems in applications of group theoretical techniques to actual problems of physical interest. In actual applications the physics of the problem not only determines the highest symmetry but also specific subgroups or subgroup chains, $G \supset H_1 \supset H_2 \supset \cdots$, where, in general, the quantum numbers associated with the irreducible representations of the physically relevant subgroups, H_n, are insufficient to give a full labelling of the irreducible representations of G. In the actual application of the Elliott $SU(3)$ symmetry to nuclear structure problems, eg, we are forced to consider the physically relevant group chain $SU(3) \supset SO(3)$, containing the exact symmetry $SO(3)$, which expresses the true rotational invariance of the physical system and the fact that the real angular momentum of the system is a bona fide good quantum number. But this subgroup chain furnishes us with only two good quantum numbers, L, M_L unlike the mathematically natural subgroup chain $SU(3) \supset SU(2) \times U(1)$ which furnishes the needed number, (three) of additional quantum numbers, through the $U(1)$ quantum number Y and the $SU(2)$ quantum numbers I, M_I. (Note that the latter involve only motion in the intrinsic x', y'-plane in the Elliott model and can thus not be associated with the orbital angular momentum of our three-dimensional world).

Another such example is the classic supermultiplet of Wigner which is the prototype of many of the higher symmetry groups used in modern particle and nuclear physics and continues to play an important role in nuclear spectroscopy. Despite this fact, there has been no complete construction of the Wigner-Racah calculus for this symmetry, (unlike the case of $SU(3)$ where computer codes, [Akiyama and Draayer, 1973],enable us to calculate any needed Wigner or Racah coefficient). This is related to the labelling problem for the $SU(4) \supset [SU(2) \times SU(2)]$ supermultiplet scheme: the spin and isospin quantum numbers, S, T, do not unambiguously identify all states in the $SU(4) \supset [SU(2) \times SU(2)]$ reduction; and consequently it is difficult to construct a group theoretically sound orthonormal basis. It is the purpose of this lecture to show how, indirectly using a complementary $Sp(6, R)$ symmetry, vector coherent state theory can lead to a group theoretically sound orthonormal $SU(4) \supset [SU(2) \times SU(2)]$ basis, [see Hecht, Le Blanc, and Rowe, 1987a]. The many earlier solutions, (some of them quite cumbersome), to the Wigner supermultiplet internal labelling problem are reviewed in this reference.

In order to understand the supermultiplet labelling problem it will be instructive to start with a specific example. For this purpose we choose the $SU(4)$ representation $[\omega] = [422]$ with a three-rowed tableau of 4, 2, and 2 squares in rows 1 to 3. Note that this is an important representation in both the nuclear $A = 12$ and $A = 24$ systems. We review first the conventional way of determining the S, T structure of such a supermultiplet, *ie* the number of occurrences of specific S, T combinations. The conventional way of carrying out the $SU(4) \rightarrow O(4) \rightarrow [SU(2) \times SU(2)]$ reduction proceeds via Littlewood's $SU(n) \rightarrow O(n)$ reduction rules. Littlewood's rule states that the possible $O(4)$ symmetries $[\lambda]$ contained in a given $SU(4)$ symmetry $[\omega] = [\omega_1\omega_2\omega_3]$ are given by the tableaux which remain after removal from the original tableau of all the possible symmetrically coupled zero-coupled pairs. The zero-coupled pair is a two-particle state coupled to $S = 0$, $T = 0$, belonging to the symmetric $SU(4)$ representation [2]. The symmetrically coupled states of n such pairs belong to $SU(4)$ representations $[n_1n_2n_3]$ with only *even* values for n_1, n_2, n_3 and $n_1+n_2+n_3 = 2n$. However, in this reduction process non-standard $O(4)$ tableaux may appear with as many as three rows. They can be converted to standard $O(4)$ tableaux $[\lambda_1\lambda_2]$ with at most two rows through the use of modification rules. For $O(4)$ the modification rules are simply

(1) $[\lambda_1\lambda_22] = -[\lambda_1\lambda_2]$

(2) $[\lambda_1 11] = [\lambda_1]$

(3) $[\lambda_1 \lambda_2 \lambda_3] = 0$ for all other cases with $\lambda_3 \neq 0$.

For the $SU(4)$ representation $[422]$ these give

Possible $[\lambda]$	(via removal of)
$\{[422](\equiv -[42])\}+$	$[0]$
$[42] + \{[321](\equiv 0)\} + \{[222](\equiv -[22])\}+$	$[2]$
$[22]+$	$[4]$
$[22] + [31] + [4]+$	$[22]$
$[2]+$	$[42]$
$[2]+$	$[222]$
$[0]$	$[422]$

Note that modification rule (1) eliminates two allowed two-rowed tableaux, (*viz* $[42]$ and one of the $[22]$), while rule (3) eliminates the tableau $[321]$. The surviving $O(4)$ symmetries of $[22] + [31] + [4] + 2[2] + [0]$ then give us the S, T structure via the $O(4) \rightarrow [SU(2) \times SU(2)]$ reduction which states simply that the S, T content of an $O(4)$ representation $[\lambda_1 \lambda_2]$ is:

$$S = \frac{1}{2}(\lambda_1 + \lambda_2), \ T = \frac{1}{2}(\lambda_1 - \lambda_2) \quad \text{and } S = \frac{1}{2}(\lambda_1 - \lambda_2), \ T = \frac{1}{2}(\lambda_1 + \lambda_2)$$

and for $[\lambda_1]$, (the special case with $\lambda_2 = 0$)

$$S = \frac{1}{2}\lambda_1, \ T = \frac{1}{2}\lambda_1 \quad \text{only.}$$

The $[422]$ supermultiplet thus contains the (S, T) values $(2,0)$, $(0,2)$, $(2,1)$, $(1,2)$, $(2,2)$, and $(0,0)$ once each, whereas $(S, T) = (1,1)$ has a double occurrence.

Our supermultiplet state construction procedure to build an orthonormal supermultiplet basis will attempt to reverse the Littlewood reduction process. Instead of removing all possible symmetrically coupled $S = 0$, $T = 0$-coupled pairs to leave states entirely free of such zero-coupled pairs, our state construction process will begin with the construction of an $O(4)$ solid harmonic of definite S and T, $Y_{ST}^{[\lambda_1\lambda_2]}$, (the "intrinsic" state), entirely free of zero-coupled pairs. This solid harmonic is then coupled with a symmetrically coupled state of n $S = 0$, $T = 0$ pairs, (the "collective" state), of symmetry $[n_1 n_2 n_3] \equiv [n]$ to make a state of good $SU(4)$ symmetry $[\omega_1\omega_2\omega_3] \equiv [\omega]$:

$$\left[Z_{00}^{[n]}(\mathbf{A}) \times Y_{ST}^{[\lambda]} \right]_{ST}^{[\omega]\varrho}.$$

Here $Z^{[n]}(\mathbf{A})$ is a polynomial of degree n in the $S = 0$, $T = 0$ pair-creation operators $\mathbf{A}$. (The multiplicity label ϱ is needed only if $[\omega]$ has a multiple occurrence in the coupling $[[\lambda] \times [n]]$).

At this stage, however, it appears from the above branching rules of our example that this construction procedure might not work. In the [422] multiplet there are two independent states with $S, T = 1, 1$, corresponding to the two possible $[n]$ values of [42] and [222] which can be combined with the solid harmonic of $O(4)$ symmetry [2] to make the [422] $SU(4)$ symmetry. However, only a *single* occurrence is predicted for the state of $O(4)$ symmetry [22], with $S, T = 2, 0$ or $0, 2$; and yet there *again* seem to be *two* ways of constructing such a state through polynomials $Z^{[n]}$ with $[n] = [4]$ *and* [22]. There appears to be an inconsistency.

However, it is clear that the above state construction does not furnish us with an orthonormal basis. It will be converted to an orthonormal basis by a K-matrix procedure. (As we shall see, the K-matrices are in fact identical with those for $Sp(6, R)$ symmetry). The K^2 overlap matrix for the $O(4)$ symmetry [22], $SU(4)$ symmetry [422],

$$\left(K^2([22], [422]) \right)_{[n][n']} = \begin{pmatrix} 4 & -2\sqrt{10} \\ -2\sqrt{10} & 10 \end{pmatrix},$$

with $[n],[n'] = [4],[22]$, in that order, contains the answer to the apparent dilemma. Its diagonal form is

$$K^2_{diag.} = \begin{pmatrix} 14 & 0 \\ 0 & 0 \end{pmatrix}.$$

One of the eigenvalues of K^2 is zero. Thus one of the two states of $O(4)$ symmetry [22] has zero norm; and there exists only a *single* state of $O(4)$ symmetry [22], corresponding to a specific linear combination of the two possible [n] values. The K^2 matrix immediately signals this important fact; and the above state construction procedure appears to be sound.

To make the construction very explicit, we introduce a set of three Bargmann 4-vectors

$$g_j^\alpha \qquad \text{with } j = 1,2,3,4; \quad \alpha = 1,2,3,$$

to work in a 12-dimensional oscillator basis. Here, $j = 1,...,4$ stand for the possible single particle spin, isospin combinations

$$|j = 1\rangle \equiv |++\rangle = \left|m_s = +\frac{1}{2}, m_t = +\frac{1}{2}\right\rangle = \frac{1}{\sqrt{2}}(|e_1\rangle - i|e_2\rangle)$$

$$|j = 2\rangle \equiv |+-\rangle = \left|m_s = +\frac{1}{2}, m_t = -\frac{1}{2}\right\rangle = \frac{1}{\sqrt{2}}(-i|e_3\rangle + |e_4\rangle)$$

$$|j = 3\rangle \equiv |-+\rangle = \left|m_s = -\frac{1}{2}, m_t = +\frac{1}{2}\right\rangle = \frac{-1}{\sqrt{2}}(i|e_3\rangle + |e_4\rangle)$$

$$|j = 4\rangle \equiv |--\rangle = \left|m_s = -\frac{1}{2}, m_t = -\frac{1}{2}\right\rangle = \frac{1}{\sqrt{2}}(|e_1\rangle + i|e_2\rangle)$$

where the e_j are 4-dimensional cartesian components. The zero-coupled pair states are then given by the scalar products

$$(\mathbf{g}^\alpha \cdot \mathbf{g}^\beta) = \sum_{i=1}^{4} g_{e_i}^\alpha g_{e_i}^\beta = A_{\alpha\beta}(\mathbf{g})$$

$$= g_{++}^\alpha g_{--}^\beta + g_{--}^\alpha g_{++}^\beta - g_{+-}^\alpha g_{-+}^\beta - g_{-+}^\alpha g_{+-}^\beta$$

$$= 2 \sum_{m_s,m_t} \langle \tfrac{1}{2} m_s \, \tfrac{1}{2} - m_s | 00 \rangle \langle \tfrac{1}{2} m_t \, \tfrac{1}{2} - m_t | 00 \rangle g_{m_s,m_t}^\alpha g_{-m_s,-m_t}^\beta,$$

where the last form shows explicitly the $S = 0$, $T = 0$ character of the pair.

The first step in our $SU(4) \supset [SU(2) \times SU(2)]$ state construction procedure involves the construction of the states entirely free of $S = 0$, $T = 0$ pair states; *ie* the generalized $O(4)$ solid harmonics which must satisfy

$$\frac{\partial^2}{(\partial \mathbf{g}^\alpha \cdot \partial \mathbf{g}^\beta)} Y_{ST}^{[\lambda]}(\mathbf{g}) = 0.$$

Note that $\partial^2/(\partial \mathbf{g}^\alpha \cdot \partial \mathbf{g}^\beta) = B_{\alpha\beta}(\mathbf{g})$ is an $S = 0$, $T = 0$-pair annihilation operator.

The simplest such solid harmonics are the totally symmetric ones, of $O(4)$ symmetry $[\lambda_1]$, which can be constructed in terms of a single Bargmann 4-vector $\mathbf{g}^1$. These must have $S = T = \frac{\lambda_1}{2}$. The state with $M_S = S$, $M_T = T$ is extremely simple

$$Y_{SS,TT}^{[\lambda_1]}(\mathbf{g}^1) = \frac{(g_{++}^1)^{\lambda_1}}{\sqrt{\lambda_1!}} \qquad \text{with } S = T = \frac{\lambda_1}{2},$$

but the general state can also be given (again with $S = T = \frac{\lambda_1}{2}$):

$$Y_{SM_S,TM_T}^{[\lambda_1]}(\mathbf{g}^1) = \left[\frac{(\frac{\lambda_1}{2} + M_S)!(\frac{\lambda_1}{2} + M_T)!(\frac{\lambda_1}{2} - M_S)!(\frac{\lambda_1}{2} - M_T)!}{\lambda_1!} \right]^{\frac{1}{2}}$$

$$\times \sum_{a=M_S+M_T}^{min(S+M_S,T+M_T)} \frac{(g_{++}^1)^a (g_{+-}^1)^{S+M_S-a} (g_{-+}^1)^{T+M_T-a} (g_{--}^1)^{a-M_S-M_T}}{a!(S+M_S-a)!(T+M_T-a)!(a-M_S-M_T)!}.$$

The general two-rowed $[\lambda_1 \lambda_2]$-solid harmonics can be written in terms of these totally symmetric functions, [Hecht, Le Blanc, and Rowe,1987a],

$$
Y^{[\lambda_1\lambda_2]}_{SM_S,TM_T} = \left[Y^{[\lambda_1]}_{\frac{\lambda_1}{2}\frac{\lambda_1}{2}}(\mathbf{g}^1) \times Y^{[\lambda_2]}_{\frac{\lambda_2}{2}\frac{\lambda_2}{2}}(\mathbf{g}^2) \right]_{SM_S,TM_T},
$$

where the square bracket denotes both spin and isospin coupling, and S, T have just the two possible values:

$$
S = \frac{1}{2}(\lambda_1 + \lambda_2), T = \frac{1}{2}(\lambda_1 - \lambda_2); \quad \text{or} \quad S = \frac{1}{2}(\lambda_1 - \lambda_2), T = \frac{1}{2}(\lambda_1 + \lambda_2).
$$

Finally, as the modification rule (2) indicates, there is a state of $O(4)$ symmetry $[\lambda_1 11]$. For this solid harmonic

$$
\begin{aligned}
Y^{[\lambda_1 11]}_{\frac{\lambda_1}{2}M_S,\frac{\lambda_1}{2}M_T} =& -\frac{1}{\sqrt{2}} \left[Y^{[\lambda_1]}_{\frac{\lambda_1}{2}\frac{\lambda_1}{2}}(\mathbf{g}^1) \times \left[Y^{[1]}_{\frac{1}{2}\frac{1}{2}}(\mathbf{g}^2) \times Y^{[1]}_{\frac{1}{2}\frac{1}{2}}(\mathbf{g}^3) \right]_{10} \right]_{\frac{\lambda_1}{2}M_S,\frac{\lambda_1}{2}M_T} \\
&+ \frac{1}{\sqrt{2}} \left[Y^{[\lambda_1]}_{\frac{\lambda_1}{2}\frac{\lambda_1}{2}}(\mathbf{g}^1) \times \left[Y^{[1]}_{\frac{1}{2}\frac{1}{2}}(\mathbf{g}^2) \times Y^{[1]}_{\frac{1}{2}\frac{1}{2}}(\mathbf{g}^3) \right]_{01} \right]_{\frac{\lambda_1}{2}M_S,\frac{\lambda_1}{2}M_T}
\end{aligned}
$$

As indicated through the modification rules, attempts to make other three-rowed $O(4)$ solid harmonics all fail; and our state construction can be made in terms of the above $Y^{[\lambda_1\lambda_2]}$ and $Y^{[\lambda_1 11]}$. Note that the above explicit expressions are those with the highest possible degree λ_1 for $\mathbf{g}^1$, and subject to this restriction the highest possible degree for $\mathbf{g}^2$, ...; ie they are extremal states for the $U(3)$ subgroup associated with the upper indices of the 12-dimensional oscillator algebra. Lower weight $U(3)$ states can be obtained from these by standard $U(3)$-lowering operations, when needed.

The construction of the symmetrically coupled polynomials of n $S = 0, T = 0$ pair operators, $\mathbf{A}$, is also straightforward. The $U(3)$ extremal states are given simply by

$$Z^{[n_1 n_2 n_3]}\big(\mathbf{A}(\mathbf{g})\big) =$$

$$N(n_1 n_2 n_3)(\mathbf{g}^1 \cdot \mathbf{g}^1)^{\frac{n_1 - n_2}{2}} \begin{vmatrix} (\mathbf{g}^1 \cdot \mathbf{g}^1) & (\mathbf{g}^1 \cdot \mathbf{g}^2) \\ (\mathbf{g}^1 \cdot \mathbf{g}^2) & (\mathbf{g}^2 \cdot \mathbf{g}^2) \end{vmatrix}^{\frac{n_2 - n_3}{2}} \begin{vmatrix} (\mathbf{g}^1 \cdot \mathbf{g}^1) & (\mathbf{g}^1 \cdot \mathbf{g}^2) & (\mathbf{g}^1 \cdot \mathbf{g}^3) \\ (\mathbf{g}^1 \cdot \mathbf{g}^2) & (\mathbf{g}^2 \cdot \mathbf{g}^2) & (\mathbf{g}^2 \cdot \mathbf{g}^3) \\ (\mathbf{g}^1 \cdot \mathbf{g}^3) & (\mathbf{g}^2 \cdot \mathbf{g}^3) & (\mathbf{g}^3 \cdot \mathbf{g}^3) \end{vmatrix}^{\frac{n_3}{2}}$$

Since the $(\mathbf{g}^\alpha \cdot \mathbf{g}^\beta)$ with $\alpha \neq \beta$ and $(\mathbf{g}^\alpha \cdot \mathbf{g}^\alpha)$ have the same Bargmann space scalar products as the $z_{\alpha\beta}$ and $z_{\alpha\alpha}$ of section 3.3, (except for a common normalization factor of 2), the normalization factor $N(n_1 n_2 n_3)$, (except for a power of 2), is that of section 3.3: $N(n_1 n_2 n_3) = N_{n_1 n_2 n_3}/2^{\frac{1}{2}(n_1 + n_2 + n_3)}$. We are thus in a position to reverse the Littlewood procedure with the construction

$$\left[Z_{00}^{[n]}(\mathbf{A}) \times Y_{ST}^{[\lambda]} \right]^{[\omega]\varrho}_{SM_S, TM_T}$$

where the coupling $[\lambda] \times [n] \to [\omega]$, (with possible multiplicity ϱ), gives the desired final $SU(4)$-symmetry $[\omega]$.

The group chain : $SU(4) \supset O(4) \supset [SU(2) \times SU(2)]$

is associated with $[\omega] \to [\lambda] \to SM_S, \, TM_T$

where S and T are related to $[\lambda]$ as indicated earlier.

Note, however, that the $A_{\alpha\beta} = (\mathbf{g}^\alpha \cdot \mathbf{g}^\beta)$, their Bargmann space hermitian conjugate partners, $B_{\alpha\beta}$, together with the upper index $U(3)$ group, satisfy the commutation relations of the $Sp(6, R)$ algebra. With $\alpha, \beta = 1, 2, 3$:

$$A_{\alpha\beta} = A_{\beta\alpha} = (\mathbf{g}^\alpha \cdot \mathbf{g}^\beta)$$

$$B_{\alpha\beta} = B_{\beta\alpha} = \frac{\partial^2}{(\partial \mathbf{g}^\alpha \cdot \partial \mathbf{g}^\beta)}$$

$$C_{\alpha\beta} = \frac{1}{2}\left((\mathbf{g}^\alpha \cdot \frac{\partial}{\partial \mathbf{g}^\beta}) + (\frac{\partial}{\partial \mathbf{g}^\beta} \cdot \mathbf{g}^\alpha) \right) = (\mathbf{g}^\alpha \cdot \frac{\partial}{\partial \mathbf{g}^\beta}) + \frac{1}{2}\delta_{\alpha\beta}4.$$

The only difference from the realization of $Sp(6, R)$ of section 3.3 in terms of $3 \times (A - 1)$ oscillator creation/annihilation operators comes from the fact that our scalar product now involves a sum over 4 spin, isospin quantum numbers instead, *ie* we have a realization in terms of 3×4 oscillator creation/annihilation operators in Bargmann space. The solid harmonic condition

$$B_{\alpha\beta} Y^{[\lambda]}_{SMs,TM_T} = 0$$

shows also that the $[\lambda]$ play the role of the "intrinsic" quantum numbers, and that the $Y^{[\lambda]}$ play the role of the generalized "vacuum" states.

The states

$$\left[Z^{[n]}_{00}(\mathbf{A}) \times Y^{[\lambda]}_{ST} \right]^{[\omega]\varrho,\eta}_{SMs,TM_T} \equiv \left| \Psi\big([\lambda];[\omega];[\mathbf{n}]\varrho;\eta\big) \right\rangle$$

thus carry not only the desired quantum numbers of the $SU(4) \supset O(4) \supset [SU(2) \times SU(2)]$ group chain but are also the (nonorthonormal) basis states of a group chain:

$$Sp(6,R) \supset U(3) \supset [SU(2) \times U(1)]$$

with quantum numbers

$$[\lambda] \longrightarrow [\omega] \longrightarrow \eta$$

(Up to now we have used only $U(3)$ extremal states so that the subgroup quantum numbers for $U(3)$, now indicated in shorthand by η, were not needed. For our present purposes the specific choice of η is again unimportant). The overlap matrix which can be used to convert the above into an orthonormal basis,

$$\big(K^2([\lambda][\omega]) \big)_{n'\varrho',n\varrho} = \big\langle \Psi\big([\lambda];[\omega];[n']\varrho';\eta\big) \big| \Psi\big([\lambda];[\omega];[n]\varrho;\eta\big) \big\rangle$$

is therefore known from the $Sp(6,R)$ algebra of section 3.3., if we make the identification $[\sigma_1 \sigma_2 \sigma_3] \rightarrow [\lambda] \equiv [\lambda_1 \lambda_2]$ or $[\lambda_1 11]$, and if we replace the quantity $A - 1$ by 4; (see *eg* the eigenvalue difference $\Lambda_{n'\omega'} - \Lambda_{n\omega}$ as given in section 3.3).

In section 3.3 we mentioned the fact that, for $A > 6$, the K^2-matrices were free of zero eigenvalues; but our initial example already illustrates the fact that, with the replacement $(A-1) \rightarrow 4$, the occurrence of zero eigenvalues must be expected and in fact becomes very important.

The equivalence between the $SU(4) \supset O(4) \supset [SU(2) \times SU(2)]$ and the complementary $Sp(6,R) \supset U(3)$ constructions can then be exploited to build a group theoretically sound orthonormal Wigner supermultiplet basis

$$|[\lambda]; [\omega]; i; SM_S, TM_T; \eta\rangle$$

$$= \sum_{n\varrho} (K^{-1}([\lambda][\omega]))_{i,n\varrho} \left[Z_{00}^{[n]}(\mathbf{A}) \times Y_{ST}^{[\lambda]} \right]_{SM_S, TM_T}^{[\omega]\varrho, \eta}$$

Due to the importance of zero eigenvalues, K^{-1} will now be chosen as

$$\left(K^{-1}([\lambda][\omega])\right)_{i,n\varrho} = \frac{1}{\sqrt{L_i}} U_{i,n\varrho},$$

where U diagonalizes K^2

$$\sum_{n\varrho} \sum_{n'\varrho'} U_{i,n\varrho} K^2_{n\varrho,n'\varrho'} U^\dagger_{n'\varrho',j} = \delta_{ij} L_i,$$

and L_i is one of the *nonzero* eigenvalues of K^2. (Recall that these were called λ_i in section 3.3; we have changed the notation to L_i to avoid confusion with the $O(4)$ quantum numbers $[\lambda]$ of this chapter).

This orthonormal basis can now be used to begin the calculation of a supermultiplet Wigner-Racah calculus. The g_j^α form the fundamental tensors of this algebra

$$g_j^\alpha \longrightarrow g_{\frac{1}{2}m_s,\frac{1}{2}m_t}^{[1]} \qquad \text{for any } \alpha,$$

and the spin, isospin-reduced matrix elements of such operators are proportional to the $SU(4)$ reduced Wigner coefficients coupling the fundamental representation to an arbitrary one

$$\left\langle \begin{array}{c} [\omega']j \\ [\lambda']S'T' \end{array} \right\| g^{[1]}_{\frac{1}{2}\frac{1}{2}} \left\| \begin{array}{c} [\omega]i \\ [\lambda]ST \end{array} \right\rangle = C([\omega][\omega']) \left\langle \begin{array}{cc} [\omega]i & [1] \\ [\lambda]ST; & \frac{1}{2}\frac{1}{2} \end{array} \right\| \left. \begin{array}{c} [\omega']j \\ [\lambda']S'T' \end{array} \right\rangle$$

where the simple spin isospin independent factor $C([\omega][\omega'])$ is given by [Hecht, Le Blanc, and Rowe, 1987a].

To evaluate such reduced matrix elements we need to know all possible products of the form

$$\left[Y^{[\lambda]}_{ST} \times g^{[1]}_{\frac{1}{2}\frac{1}{2}} \right]^{[k]}_{S'T'}$$

where the square bracket denotes both the $[\lambda] \times [1] \rightarrow [k]$ coupling in the upper quantum numbers and spin and isospin coupling in the lower S, T quantum numbers. There are two possibilities:

1) With $[k] = [\lambda']$

$$\left[Y^{\lambda}_{ST} \times g^{[1]}_{\frac{1}{2}\frac{1}{2}} \right]^{[\lambda']}_{S'T'} = F([\lambda], [\lambda']) Y^{[\lambda']}_{S'T'}$$

2)

$$\left[Y^{\lambda}_{ST} \times g^{[1]}_{\frac{1}{2}\frac{1}{2}} \right]^{[k]}_{S'T'} = G([\lambda], [\lambda']; [k]) \left[Z^{[2]}_{00} \times Y^{[\lambda']}_{S'T'} \right]^{[k]}_{S'T'}.$$

Examples are

$$\left[Y^{[\lambda_1 \lambda_2]}_{(\frac{\lambda_1}{2} + \frac{\lambda_2}{2}),(\frac{\lambda_1}{2} - \frac{\lambda_2}{2})} \times g^{[1]}_{\frac{1}{2}\frac{1}{2}} \right]^{[\lambda_1 + 1, \lambda_2]}_{(\frac{\lambda_1}{2} + \frac{\lambda_2}{2} + \frac{1}{2}),(\frac{\lambda_1}{2} - \frac{\lambda_2}{2} + \frac{1}{2})}$$

$$= \sqrt{\frac{(\lambda_1 + 2)(\lambda_1 - \lambda_2 + 1)}{(\lambda_1 - \lambda_2 + 2)}} Y^{[\lambda_1 + 1, \lambda_2]}_{(\frac{\lambda_1}{2} + \frac{\lambda_2}{2} + \frac{1}{2}),(\frac{\lambda_1}{2} - \frac{\lambda_2}{2} + \frac{1}{2})}$$

and

$$\left[Y^{[\lambda_1 \lambda_2]}_{(\frac{\lambda_1}{2} + \frac{\lambda_2}{2}),(\frac{\lambda_1}{2} - \frac{\lambda_2}{2})} \times g^{[1]}_{\frac{1}{2}\frac{1}{2}} \right]^{[\lambda_1 + 1, \lambda_2]}_{(\frac{\lambda_1}{2} + \frac{\lambda_2}{2} - \frac{1}{2}),(\frac{\lambda_1}{2} - \frac{\lambda_2}{2} - \frac{1}{2})}$$

$$= \sqrt{\frac{(\lambda_1 + \lambda_2 + 1)}{2(\lambda_1 + \lambda_2)(\lambda_1 + 1)}} \left[Z_{00}^{[2]} \times Y_{(\frac{\lambda_1}{2}+\frac{\lambda_2}{2}-\frac{1}{2}),(\frac{\lambda_1}{2}-\frac{\lambda_2}{2}-\frac{1}{2})}^{[\lambda_1-1,\lambda_2]} \right]_{(\frac{\lambda_1}{2}+\frac{\lambda_2}{2}-\frac{1}{2}),(\frac{\lambda_1}{2}-\frac{\lambda_2}{2}-\frac{1}{2})}^{[\lambda_1+1,\lambda_2]}$$

Altogether there are 4 possibilities of type 1) and 10 possible cases of type 2). The simple functions $F([\lambda],[\lambda'])$, $G([\lambda],[\lambda'];[k])$ for all cases are listed by [Hecht, Le Blanc, and Rowe, 1987a]. With these results and our $Sp(6,R) \supset U(3)$ vector coherent state construction it is straightforward to calculate the spin, isospin reduced matrix element of a fundamental tensor: (Recall our right to left coupling order convention).

$$\left\langle \begin{matrix} [\omega']j \\ [\lambda']S'T' \end{matrix} \middle\| g_{\frac{1}{2}\frac{1}{2}}^{[1]} \middle\| \begin{matrix} [\omega]i \\ [\lambda]ST \end{matrix} \right\rangle$$

$$= \sum_{n\varrho} (K^{-1}([\lambda][\omega]))_{i,n\varrho} \left\langle \begin{matrix} [\omega']j \\ [\lambda']S'M_S'T'M_T' \end{matrix} \middle\| \left[\left[Z_{00}^{[n]} \times Y_{ST}^{[\lambda]} \right]_{ST}^{[\omega]\varrho} \times g_{\frac{1}{2}\frac{1}{2}}^{[1]} \right]_{S'M_S'T'M_T'}^{[\omega']} \right\rangle$$

$$= \sum_{n\varrho} (K^{-1}([\lambda][\omega]))_{i,n\varrho} \sum_{[k]\varrho_k} U([1][\lambda][\omega'][n]; [k]_{-\varrho_k}; [\omega]\varrho_-)$$

$$\times \left\langle \begin{matrix} [\omega']j \\ [\lambda']S'M_S'T'M_T' \end{matrix} \middle\| \left[Z_{00}^{[n]} \times \left[Y_{ST}^{[\lambda]} \times g_{\frac{1}{2}\frac{1}{2}}^{[1]} \right]_{S'T'}^{[k]} \right]_{S'M_S'T'M_T'}^{[\omega']\varrho_k} \right\rangle.$$

In the case of category 1) above, where $[Y^{[\lambda]} \times g^{[1]}]$ collapses to the simple $Y^{[\lambda']}$, this yields

$$\left\langle \begin{matrix} [\omega']j \\ [\lambda']S'T' \end{matrix} \middle\| g_{\frac{1}{2}\frac{1}{2}}^{[1]} \middle\| \begin{matrix} [\omega]i \\ [\lambda]ST \end{matrix} \right\rangle$$

$$= \sum_{n\varrho,\varrho'} (K^{-1}([\lambda][\omega]))_{i,n\varrho} U([1][\lambda][\omega'][n]; [\lambda']_{-\varrho'}; [\omega]\varrho_-)$$

$$\times F([\lambda],[\lambda']) (K([\lambda'][\omega']))_{n\varrho',j}.$$

In the case of category 2) an additional recoupling is necessary to combine $Z^{[n]}$

with $Z^{[2]}$, and yields

$$\left\langle \begin{matrix} [\omega']j \\ [\lambda']S'T' \end{matrix} \left\| g^{[1]}_{\frac{1}{2}\frac{1}{2}} \right\| \begin{matrix} [\omega]i \\ [\lambda]ST \end{matrix} \right\rangle$$

$$= \sum_{n\varrho} \sum_{n'\varrho'} \sum_{[k]\varrho_k} (K^{-1}([\lambda][\omega]))_{i,n\varrho} U([1][\lambda][\omega'][n]; [k]_{-}\varrho_k; [\omega]\varrho_{-})$$

$$\times U([\lambda'][2][\omega'][n]; [k]_{-}\varrho_k; [n']_{-}\varrho') G([\lambda], [\lambda']; [k])$$

$$\times \langle [n'\|\mathbf{z}\|[n]\rangle (K([\lambda'][\omega']))_{n'\varrho',j}.$$

We have also used the fact that the $Z(\mathbf{A})$ satisfy

$$\left[Z^{[n]}_{00} \times Z^{[2]}_{00} \right]^{[n']}_{00} = Z^{[n']}_{00} \langle [n']\|\mathbf{z}\|[n]\rangle,$$

where the reduced matrix element is that of section 3.3; the purely "collective" reduced matrix element of the 6-dimensional collective variable $\mathbf{z}$ reduced with respect to $U(3)$.

The Racah coefficients in these relations are $U(3)$-recoupling coefficients and are readily available through the code of [Akiyama and Draayer, 1973]. The basic results then bear this similarity to those of the last chapter. The $SU(4)$ Wigner coefficients are given by simple coefficients, such as the $Sp(6, R)$ K-matrix elements, or simple functions such as the $F([\lambda], [\lambda'])$, and the recoupling (Racah) coefficients of a simpler symmetry group, in this case $U(3)$. Thus $SU(4)$ Wigner coefficients can be evaluated from a knowledge of $U(3)$ Racah coefficients.

The example of this chapter is by no means an isolated example of the applicability of vector coherent state theory. In fact it was based on the earlier analysis of [Le Blanc and Rowe, 1985 a and b] of the $SU(3) \supset SO(3)$ symmetry mentioned in the introduction. In this case the analysis is made in terms of two

Bargmann 3-vectors

$$(g_x^\alpha, g_y^\alpha, g_z^\alpha) \qquad \text{with} \quad \alpha = 1, 2$$

where x, y, z refer to our 3-dimensional world. The zero-coupled pairs in this case are pairs, $(\mathbf{g}^\alpha \cdot \mathbf{g}^\beta) = A_{\alpha\beta}$, coupled to orbital angular momentum, $L = 0$. The $O(3)$ solid harmonics, states entirely free of $L = 0$ pairs, are in this case of two kinds: (1) totally symmetric functions, $Y_{LM_L}^{[\lambda]}(\mathbf{g}^1)$ with $\lambda = L$, standard 3-dimensional solid harmonics; and (2), bearing in mind the $O(3)$ modification rule $[\lambda 1] \equiv [\lambda]$,

$$Y_{LM_L}^{[L1]} = -\sqrt{\frac{2}{L+1}} [Y_{L-1}^{[L-1]}(\mathbf{g}^1) \times [Y_1^{[1]}(\mathbf{g}^1) \times Y_1^{[1]}(\mathbf{g}^2)]_1^{[11]}]_{LM_L}^{[L1]}.$$

The n symmetrically coupled $L = 0$ pairs give rise to functions of $[n_1 n_2]$ symmetry with n_1 and n_2 both even, $n_1 + n_2 = 2n$. The $SU(3) \supset SO(3)$ state construction is achieved through

$$\left[Z_0^{[n_1 n_2]}(\mathbf{A}) \times Y_L^{[Le]} \right]_{LM_L}^{[\omega_1 \omega_2]\eta},$$

where $[Le]$, $e = 0$ or 1, gives the $O(3)$ symmetry, and the $SU(3)$ irreducible representation is specified by the two-rowed tableau $[\omega_1 \omega_2]$. However, these functions can be classified not only according to the $SU(3) \supset O(3) \supset SO(3)$ chain; but also according to the complementary symmetry $Sp(4, R) \supset U(2)$ where the $Sp(4, R)$ quantum numbers are $[Le]$ and the $U(2)$ quantum numbers are $[\omega_1 \omega_2]$. Now the above construction can be converted into an orthonormal basis through the $Sp(4, R)$ K^2-matrices.

References

Akiyama Y. and Draayer J.P.; 1973
  Comput. Phys. Commun. **5** 405 and J.Math. Phys. **14** 1904

Arechi F.T., Courtens E., Gilmore R., and Thomas H.; 1972
  Phys. Rev. **A6** 2211

Barut A.O. and Girardello L.; 1971
  Commun. Math. Phys. **21** 41

Castaños O., Chacón E., and Moshinsky M.; 1984
  J.Math. Phys. **25** 1211

Castaños O., Chacón E., Moshinsky M., and Quesne C.; 1985
  J. Math. Phys. **26** 2107

Castaños O., Kramer P., and Moshinsky M.; 1985
  J. Math. Phys. **27** 924

Casten R.F., Wu Cheng-Li, Feng Da Hsuan, Ginocchio J.N., and Han Xiao-Ling;
  1986
  Phys. Rev. Lett. **56** 2578

Chen Jin-Quan, Feng Da Hsuan, and Wu Cheng-Li; 1986
  Phys. Rev. **C34** 2269

Deenen J. and Quesne C.; 1984a,b
  J.Math. Phys. **25** 1638, 2354

Deenen J. and Quesne C.; 1984c
  J.Phys. A: Math. Gen. **17** L404

Dobacewski J.; 1981
  Nucl. Phys. **A369** 213, 237

Dobacewski J.; 1982
  Nucl. Phys. **A380** 1

Gilmore R.; 1974

"Lie Groups, Lie Algebras, and Some of their Applications" (Wiley, New York)

Ginocchio J.N.; 1980

Ann. Phys. **126** 234

Glauber R.J.; 1963a,b

Phys. Rev. **130** 2529 and **131** 2766

Hecht K.T.; 1985a

Nucl. Phys. **A444** 189

Hecht K.T.; 1985b

Proc. of VIIIth Oaxtepec Nuclear Physics Symposium (UNAM, Mexico)

Hecht K.T.; 1987 Preprint

"Vector Coherent State Theory for the S , D Fermion Pair Algebra" (to be publ in Nucl. Phys. A)

Hecht K.T. and Elliott J.P.; 1985

Nucl. Phys. **A438** 29

Hecht K.T., Le Blanc R., and Rowe D.J.; 1987a

J.Phys. A: Math. Gen. **20** 257

Hecht K.T., Le Blanc R., and Rowe D.J.; 1987b

J.Phys. A: Math. Gen. **20** 2241

Klauder J.R. and Skagerstam B.S.; 1985

"Coherent States" (World Scientific, Singapore)

Le Blanc R.; 1987

J.Phys. A: Math. Gen. **20** (in press)

Le Blanc R. and Hecht K.T.; 1987

J.Phys. A: Math. Gen. **20** (in press)

Le Blanc R. and Rowe D.J.; 1985a,b

J.Phys. A: Math. Gen. **18** 1891, 1905

Le Blanc R. and Rowe D.J.; 1986a,b
 J.Phys. A: Math. Gen. **19** 1093, 1111

Le Blanc R. and Rowe D.J.; 1987a
 J. Math. Phys. **28** 1231

Le Blanc R. and Rowe D.J.; 1987b
 J.Phys. A: Math. Gen. **20** L681

Le Blanc R. and Rowe D.J.; 1987 Preprint a
 "The Matrix Representation of g_2. I. Representations in an $so(4)$ Basis."

Le Blanc R. and Rowe D.J.; 1987 Preprint b
 "The Matrix Representations of g_2. II. Representations in an $su(3)$ Basis."

Louck J.D.; 1970
 American J. Phys. **38** 3

Perelomov A.; 1986
 "Generalized Coherent States and Their Applications"
 Texts and Monogr. in Physics (Springer, Berlin, Heidelberg)

Quesne C.; 1981
 J.Math. Phys. **22** 1482

Quesne C.; 1986a,b
 J. Math. Phys. **27** 428, 869

Rowe D.J.; 1984a
 J.Math. Phys. **25** 2662

Rowe D.J.; 1984b
 "Coherent States, Contractions and Classical Limits of the Noncompact Symplectic Groups" in Proc. XIIIth Int. Colloq. on Group Theor. Methods in Phys., College Park, Md., ed. W.W. Zachary (World Scientific, Singapore)

Rowe D.J.; 1985
 Rep. Prog. Phys. **48** 1419

Rowe D.J.; 1986

"Some Recent Advances in Coherent State Theory and its Applications to Nuclear Collective Motion" in Phase Space Approach to Nuclear Dynamics, di Toro, Nörenberg, Rosina, and Stringari, eds. (World Scientific, Singapore) p. 546

Rowe D.J. and Carvalho J., 1986
Phys. Lett. **175** 243

Rowe D.J., Le Blanc R., and Hecht K.T.; 1987 Preprint
"Vector Coherent State Theory and its Application to the Orthogonal Groups" (to be publ. in J.Math. Phys.)

Rowe D.J., Rosensteel G., and Carr R.; 1984
J.Phys. A: Math. Gen. **17** L399

Rowe D.J., Rosensteel G., and Gilmore R.; 1985
J.Math. Phys. **26** 2787

Rowe D.J., Wybourne B.G., and Butler P. 1985
J.Phys. A: Math. Gen. **18** 939

Vergados J.D.; 1968
Nucl. Phys. **111** 681

Vogel P. and Zirnbauer M.; 1986
Phys. Rev. Lett. **57** 3148

Wu Cheng-Li, Feng Da Hsuan, Chen Xuan-Gen, Chen Jin-Quan, and Guidry M.W.; 1986
Phys. Lett. **168B** 313